INSIDE ENERGY

Developing and Managing an ISO 50001
Energy Management System

Charles H. Eccleston
Frederic March
Timothy Cohen

CRC Press
Taylor & Francis Group
Boca Raton London New York

CRC Press is an imprint of the
Taylor & Francis Group, an **informa** business

CRC Press
Taylor & Francis Group
6000 Broken Sound Parkway NW, Suite 300
Boca Raton, FL 33487-2742

© 2012 by Taylor & Francis Group, LLC
CRC Press is an imprint of Taylor & Francis Group, an Informa business

No claim to original U.S. Government works

Printed in the United States of America on acid-free paper
Version Date: 20111004

International Standard Book Number: 978-1-4398-7670-1 (Hardback)

This book contains information obtained from authentic and highly regarded sources. Reasonable efforts have been made to publish reliable data and information, but the author and publisher cannot assume responsibility for the validity of all materials or the consequences of their use. The authors and publishers have attempted to trace the copyright holders of all material reproduced in this publication and apologize to copyright holders if permission to publish in this form has not been obtained. If any copyright material has not been acknowledged please write and let us know so we may rectify in any future reprint.

Except as permitted under U.S. Copyright Law, no part of this book may be reprinted, reproduced, transmitted, or utilized in any form by any electronic, mechanical, or other means, now known or hereafter invented, including photocopying, microfilming, and recording, or in any information storage or retrieval system, without written permission from the publishers.

For permission to photocopy or use material electronically from this work, please access www.copyright.com (http://www.copyright.com/) or contact the Copyright Clearance Center, Inc. (CCC), 222 Rosewood Drive, Danvers, MA 01923, 978-750-8400. CCC is a not-for-profit organization that provides licenses and registration for a variety of users. For organizations that have been granted a photocopy license by the CCC, a separate system of payment has been arranged.

Trademark Notice: Product or corporate names may be trademarks or registered trademarks, and are used only for identification and explanation without intent to infringe.

Visit the Taylor & Francis Web site at
http://www.taylorandfrancis.com

and the CRC Press Web site at
http://www.crcpress.com

Contents

Foreword ix
Preface xi
The Authors xix
Introduction xxi

**Chapter 1 General requirements for an energy
            management system** 1
1.1   Qualifications 1
1.2   Requirements for organizations 2
1.3   EnMS requirements 3

Chapter 2 Management responsibility 5
2.1   General requirements 5
2.2   Establish, implement, and maintain the energy policy 7
2.3   Management representative and energy management team 8
2.4   Provide energy management system resources 9
2.5   Defining the energy management system scope
      and boundaries 10
2.6   Communicating the importance of energy management 12
2.7   Establishing energy performance objectives and targets 13
2.8   Energy performance indicators 15
2.9   Including energy considerations in long-term planning 16
2.10 Ensure that results are measured and reported 17
2.11 Conduct management reviews 18
2.12 Roles, responsibility, and authority 19
2.13 Establish, implement, maintain, and continually improve the
      energy management system 20
2.14 Report to top management on the performance of the energy
      management system 21
2.15 Report to top management on changes in energy
      performance 22
2.16 Forming the energy management team 23

2.17 Plan and direct energy management activities designed to support the organization's energy policy 24
2.18 Defining and communicating responsibilities and authorities 25
2.19 Determining criteria and methods for effective operation and control of the energy management system 26

Chapter 3 Energy policy 29
3.1 Commitment to improvement 29
3.2 Criteria 30

Chapter 4 Energy planning 37
4.1 General 37
4.2 Legal and other requirements 40
4.3 Energy review 42
4.4 Energy baseline 48
4.5 Energy performance indicators 51
4.6 Objectives, targets, and action plans 53

Chapter 5 Implementation and operation 59
5.1 General 59
5.2 Competence, training, and awareness 60
5.3 Awareness of the energy policy, the energy management system, and procedures 61
5.4 Roles, responsibilities, and authorities 62
5.5 Benefits of improved energy performance 63
5.6 Individual impacts and contributions to improved energy performance 64
5.7 Documentation requirements 65
5.8 Documenting scope and boundaries 66
5.9 Documenting the energy policy 67
5.10 Documenting energy objectives, targets, and action plans 68
5.11 Maintaining documents considered by the organization to be necessary for ensuring planning, operation, and control 68
5.12 Control of documents 69
5.13 Approving documents prior to issue 72
5.14 Periodic review and update 72
5.15 Identifying changes and current revision status 73
5.16 Ensuring that relevant versions of applicable documents are available at points of use 74
5.17 Ensuring documents are legible and readily identifiable 75
5.18 Controlling external documents 76
5.19 Preventing the unintended use of obsolete documents 77
5.20 Operational control 78

Contents   v

5.21 Effectively operating and maintaining significant energy use 81
5.22 Operating and maintaining facilities, processes, systems, and equipment 82
5.23 Communicating operational controls to personnel 82
5.24 Communications 83
5.25 Designing facilities, equipment, systems, and processes 90
5.26 Procurement of energy services, products, and equipment 92
5.27 Assessing products' energy use over time 93
5.28 Consider contingency and emergency situations and potential disasters 94
5.29 Procurement of energy supply 95

Chapter 6 Checking performance 97
6.1 Monitoring, measurement, and analysis 97
6.2 Evaluation of legal and other compliance 99
6.3 Internal audit of the EnMS 100
6.4 Nonconformities, correction, corrective, and preventive action 104
6.5 Reviewing and determining the causes of nonconformities and potential nonconformities 107
6.6 Evaluating the need for action 108
6.7 Implementing corrective and preventive actions 111
6.8 Maintaining records of corrective and preventive actions 112
6.9 Reviewing the effectiveness of the corrective or preventive action taken 113
6.10 Control of records 114
Endnote 116

Chapter 7 Management review 117
7.1 Summary of requirements 117
7.2 Input to management review 118
7.3 Actions from previous management reviews 119
7.4 Energy policy 120
7.5 Energy performance and energy performance indicators 120
7.6 Legal compliance and requirement changes 120
7.7 Energy objectives and targets 120
7.8 Energy management system audit results 121
7.9 Corrective and preventive actions 121
7.10 Projected energy performance 122
7.11 Recommendations for improvement 122
7.12 Output from management review 123
7.13 Changes in the energy performance of the organization 123
7.14 Changes to the energy policy 124

7.15	Changes to the EnPIS	124
7.16	Continual improvement of the energy management system and its implementation	124
7.17	Allocation of resources	125
Bibliography		127

Appendix A: Energy consumption, generation, sustainability, and energy systems 129
- A.1 Energy consumption 129
- A.2 Energy production 134
- A.3 Sustainability 138
- A.4 A survey of energy sources 139
- Endnotes 153

Appendix B: Perspectives on energy efficiency and conservation 155
- B.1 Energy conservation 156
- B.2 Energy use and efficiency around the world 158
- B.3 Energy conservation/energy efficiency issues 160
- Endnotes 161

Appendix C: Peak oil: The looming oil crisis 163
- C.1 King Hubbert's prediction 164
- C.2 Growing oil consumption and lower exports by producers 168
- C.3 Peak oil policy implications 169
- C.4 Concluding thoughts 176
- Endnotes 177

Appendix D: Sustainability and energy policy 179
- D.1 Definitions of sustainability 179
- D.2 Sustainability impact 182
- D.3 Agenda 21 184
- D.4 Common principles of sustainability 184
- D.5 Measuring sustainable development 189
- D.6 Applications of sustainable resource development 193
- D.7 Examples of a sustainable energy policy 196
- Endnotes 203

Appendix E: Global climate change 207
- E.1 The nature of the problem 207
- E.2 Global warming causal factors and research 211
- E.3 Current and future impacts of global warming 219
- E.4 Reducing greenhouse emissions 223
- E.5 Global climate policy 228
- Endnotes 238

Appendix F: Selected key ISO 50001 definitions 243

Appendix G: The energy assessment 245
G.1 The assessment process 247
G.2 Performing the energy assessment 250
Endnotes 262

Appendix H: Methods, tools, and techniques 265

Appendix I: Applying the nominal group technique to maximize
return on investment 269
I.1 Typical NGT procedure 270
I.2 Advantages and disadvantages 271
Endnotes 271

Appendix J: The Fukushima Daiichi nuclear power plant disaster 273

Index 281

Foreword

The world's first energy crisis

It was a progressive crisis, one that had no parallel in the annals of history. There was no distinct event that actually marked it, although it was clearly beginning to fester around the year 1500 AD. This crisis only intensified as demand increasingly outstripped supply. Even so, it was not initially a universal problem, but one that started principally in England and crept outward to other areas of Europe. Before it was over, it would come to jeopardize the very advancement of European civilization.

The demand for wood—which heated houses and cooked meals, was the lifeblood of ironworking, and provided the building material for the great sailing ships of the time—soared to the point that it could no longer be supplied from within the British Isles. The price of this commodity began to rise sharply. Unsustainable wood harvests had decimated the forests of England and, to a lesser extent, a significant portion of Europe as well. Most of Europe found itself in the midst of a wood crisis by the end of the 17th century. Eventually, new wooden buildings were banned in London. The long era in which wood had fueled Western civilization was coming to a screeching halt. The world's first energy crisis was at hand.

The world's first energy crisis was now threatening the Industrial Revolution and the reemergence of Western society. Western civilization was flirting perilously close to a return to the Dark Ages.

Coal deposits, which occasionally outcropped along the land surface, had for centuries been more of a curiosity than a widespread source of energy. This soft black rock would save everything. Inventors of the day learned how to harness the fantastic power of coal in the nick of time. The era of coal was born. They learned to heat homes, cook food, manufacture iron, and run machines. Coal powered the Industrial Revolution. James Watt would go on to invent the steam engine, which was fired by this shiny black rock. It would fast-forward the Industrial Revolution. Coal soon made its way to the United States. The rest was history.

A crisis across Europe had been averted. But the transition from wood to coal would not be an easy one. Coal was difficult and hazardous to mine. It was dirty. The skies in London were turned dark with smog and soot.

In the nineteenth century, coal began to be replaced by oil, an even more powerful energy source, and still later by nuclear energy. But recall that Europe's brush with disaster had been the result of placing unsustainable demands on its timber resources. Today, alarms are being issued that everything from oil, to minerals, to our agricultural industry (which is predicated on cheap and abundant energy) are being utilized at unsustainable rates. Will we dodge another disaster by blind luck, much as Europe did with its timber industry, or is another day coming such as the recent disaster at Fukushima (Appendix J) when we will indeed have to pay the piper? It is for reasons such as this that this book has been written.

Preface

This book is designed to assist high-level organizational managers and their energy systems teams through the project and program steps that can transform existing energy management systems to far more effective ones that significantly reduce the costs of energy in a business's bottom line.

The International Standards Organization has recently promulgated "ISO 50001 Energy Management Systems—Requirements With Guidance For Use," which provides a systematic and comprehensive approach grounded on the well-established principles of its ISO 9001 Quality Management Systems series that has been adopted by thousands of private and public sector organizations worldwide. Our approach is twofold:

1. In Chapters 1–7 we help the user understand and apply each of the many requirements of the standard in a systematic and comprehensive manner, informed by the authors' extensive experience in helping organizations improve their management systems' performance. The intent is to help the user transform existing suboptimal energy practices into a state-of-the-art, high-quality system that produces a demonstrably high return on the investment made. This outcome is assured by building the necessary processes into the management system.
2. In the various appendices we provide perspectives on multinational and national energy and environment policies that will likely affect the cost of energy purchased in the world's markets. We also offer additional guidance about methods available to management and energy teams when implementing the ISO 50001 requirements

This book reflects its authors' collective experience in developing ISO compliant management systems. The lead author (Charles Eccleston) was an elected member to the ISO/PC 242 working group, whose mission was to provide American input in the development of the ISO 50001 Energy Management System (EnMS) standard, and who has published papers offering new approaches for implementing ISO 14001 environmental management systems. Author Tim Cohen is the project manager responsible for leading Sandia National Laboratories through two large ISO management system certifications, and author Frederic March was his principal systems analyst on these projects. Please consider the guidance provided by these authors in light of the specific situation of your own organization. As you will learn herein, decisions about your energy system will require ongoing technical analysis to help you to sufficiently understand the system's performance in making sound decisions. The book seeks to help you manage the skills, knowledge, and experience of the many experts who will participate in your organization's EnMS policy planning and implementation.

How this book is organized

Except for this introduction and the book's appendices, each numbered chapter corresponds to the major seven sections of ISO 50001 Chapter 4, Energy Management Requirements as follows:

Book chapter	ISO 50001 section
1: General Requirements	4.1
2: Management Responsibility	4.2
3: Energy Policy	4.3
4: Energy Planning	4.4
5: Implementation and Operation	4.5
6: Checking Performance	4.6
7: Management Review	4.7
Appendix A: Energy Consumption, Generation, Sustainability, and Energy Systems	
Appendix B: Perspectives on Energy Efficiency and Conservation	
Appendix C: Peak Oil: The Looming Oil Crisis	
Appendix D: Sustainability and Energy Policy	
Appendix E: Global Climate Change	
Appendix F: Selected Key ISO 50001 Definitions	
Appendix G: The Energy Assessment	
Appendix H: Methods, Tools, and Techniques for Implementing ISO 50001 EnMS	
Appendix I: Applying the Nominal Group Technique to Maximize Return on Investment	
Appendix J: Fukushima Daiichi Nuclear Power Plant Disaster	

Appendices A–E provide global perspectives on energy and environmental issues likely to affect the future costs of energy. Appendices F–I offer practical information that further informs management and technical decision making in applying ISO 50001. Appendix J is a case study on the Fukushima Daiichi nuclear power plant disaster.

The ISO 50001 standard also includes a foreword, preface, and introduction, plus the following preliminary numbered chapters that are about the standard itself and do not establish specific requirements to be implemented:

Chapter 1: Scope
Chapter 2: Normative References
Chapter 3: Terms and Definitions

Figure 1 shows the seven chapters of this book that correspond to each major requirement section of the ISO 50001 standard. We have organized our guidance to each of the chapter sections to answer the following questions:

- What does this mean?
- What are the benefits?
- How do we implement?
- What evidence is needed

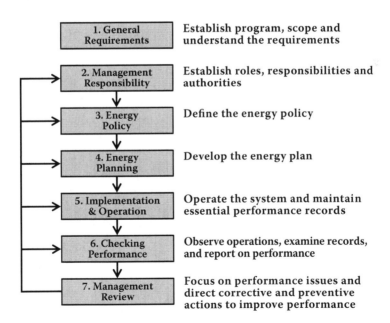

Figure 1 ISO 50001 flowchart.

Since ISO standards are subject to periodic revision, it is essential for users to obtain the most current published version of ISO 50001 when using this book for guidance.

Objectives

Our objective is to enable an organization's management to achieve its own energy performance targets and to thereby qualify for ISO 50001 registration. While there are a number of distinct benefits in implementing an ISO 50001 EnMS (see Introduction), the three principle ones are to

1. Save the organization money by increasing energy efficiency and/or reducing and more effectively managing energy generation or usage
2. Reduce generation of greenhouse gas (GHG) emissions that the majority of the scientific community believes is the principal driver behind global climate change
3 Promote improved public relations by demonstrating that the organization is making measurable and tangible efforts (ISO 50001) to manage energy

About ISO and your business or organization

Why is energy management important to my business?

Commercially available energy in all its forms represents a sizable chunk of organizational business costs for indoor climate control in buildings, transporting goods and people, and in the manufacturing processes. It is also embedded in the costs of virtually everything we purchase. A large portion of this energy budget pays directly for fossil fuels (derived from petroleum, coal, shale, and natural gas) and is indirectly embedded in the prices we pay for items essential to run our businesses. Any business owner understands that stable fossil fuel prices are essential to sustainably profitable operations.

Many business owners, large and small, are concerned with the history of increasing fossil fuel prices, and with the global and domestic factors that threaten to drive prices much higher in the coming decade. The question for most energy economists is not "will it happen?" but what the timing will be.

Fortunately there are common sense steps that a business can take to greatly reduce its dependency on fossil fuels. But they do involve financial risks because they require investments in alternative energy and conservation technologies that need to be assessed in terms of costs versus benefits, payback periods, and rates of return.

Decision outcomes from these assessments depend on the future assumed prices of fossil fuel products in a given analysis. We advise that such analysis cover a range of future energy prices. This will enable management to assess the relative risks versus returns of alternative energy investment strategies.

Many organizations that have assessed their energy costs and business risks have already begun to invest in transforming the ways in which they manage their energy.

How will ISO 50001 help me reduce my energy costs?

ISO 50001 was designed as a common-sense management template to guide organizations to significantly reduce their energy costs through prudent investments, coupled with wise implementation of energy systems. The philosophy and approach of ISO 50001 are consistent with previous ISO standards such as the ISO 9000 Quality Management Systems and ISO 14000 Environmental Management Systems. These guides help the organization to greatly improve how they manage various aspects of their business. However, the guidance must be augmented by informed and capable managers who provide clear and confident leadership tailored to the specific needs of each business.

Appendix C (Peak oil) is designed to help you develop your own perspective on this issue because of its great importance to the decisions that today's business leaders must make to protect their future.*

Why do we believe that global climate change (Appendix E) should be considered in business decisions about energy management?

Policies that ignore global climate change along with peak oil include a legacy of governmental incentives that stimulate continuing discovery and exploitation of fossil energy resources to satisfy the needs of growing world economies. Such policies implicitly assume that fossil fuel supplies will always be sustainable.

But there is a caveat. Many nations have already adopted policies in response to scientific claims that generation of greenhouse gases may be causing warming of the Earth's climate.

Policies driven by concerns for global warming include reducing or eliminating past subsidies to energy extraction industries, taxing greenhouse gas emissions, and various permit schemes that allow a fixed limit on emissions to be monitored for compliance and to allow companies to buy

---

* Appendix C was previously published as Chapter 11: Peak Oil—The Looming World Crisis by Eccleston and March in *Global Environmental Policy*. CRC Press, December 2010.

and sell permits at an exchange. Other policies designed to stimulate alternatives to fossil fuels involve incentives to investment in renewable energy technologies, and a panoply of simple and complex conservation measures.

What would the effects of global climate change policies mean to my organization?

The net effect of these policies, if they continue to attract global governments, would clearly be to increase the market costs of fossil energy while reducing the costs of renewable and conservation energy technologies to businesses and consumers. In fact, many new enterprises have emerged in the last several years that see these policies as enabling these technologies to grow large markets.

If your company is a large fossil energy consumer, it would mean that the costs of your current energy use may impede your competitiveness. If you are a fossil energy–producing company, it would mean increasing costs of managing extraction from deeper depths in oceans and on land, which would add to unit fuel costs. However, like some oil companies have already done, you would consider diversifying your business to provide alternative energy options.

Given the political controversies that cloud the climate change issue it is difficult to predict if, when, or how policies to retard global warming will likely affect the costs of fossil fuels to your organization. To help you develop your own perspective on the potential effects of this issue we have included Appendix E, Global climate change.

What else should I know that could influence decisions about our current energy management?

We believe that that you need some additional perspective on the global issues that could affect energy supply, demand, and price as well as some basic statistics about trends in global energy production, including some of the inroads being made by companies selling renewable and conservation technologies and services. Appendix A, Energy consumption, generation, sustainability, and energy systems, is designed to introduce a big picture vision of the global energy situation and to stimulate further thought and research into near or long-term opportunities for your company's energy system.

How should all these considerations affect how I apply ISO 50001?

They will help you make fundamental cost-effective decisions on when and how to invest in replacing energy system components with new technologies and energy sources aimed at reducing energy system operating costs.

Audience

This book is designed to instruct all team members contributing to the development and implementation of an organization's ISO 50001 EnMS processes. It provides practical instruction on how to design a policy, develop a plan, take appropriate actions pursuant to the plan, and monitor the performance of the EnMS to achieve continuous improvement. A list of those who will find this book useful, includes, but is not limited to

- Owners and managers of any business whose operating costs include energy expenditures to light, ventilate, heat, and cool buildings; to operate manufacturing systems; and to operate fleets of vehicles for moving materials and people
- Owners and managers of consulting firms who perform energy audits, design and install energy management systems, and train people to manage and operate them
- Elected officials and infrastructure managers for communities and municipalities who manage buildings, transportation, water resources and other energy consuming infrastructure.
- Federal and state infrastructure executives and managers
- Federal and state regulatory personnel
- Nonprofit organizations
- Power production and distribution companies

Those wishing specialized training in ISO 50001 or professional assistance in developing and implementing such a system may contact the Charles Eccleston (EcoTraining@ymail.com) or Frederic March (fmarch@thinkwellassociates.com).

The Authors

Charles H. Eccleston is an author, trainer, and environmental/energy consultant. He is recognized in Marquis' *Who's Who in America* and *Who's Who in the World* as a leading international expert for his innovative environmental policy and environmental impact assessment (EIA) achievements. With over 20 years of experience, he has managed and prepared numerous environmental and energy assessments and policy studies. Eccleston is the author of over 60 professional papers and eight books on the U.S. National Environmental Policy Act (NEPA) process, EIA, and environmental/ energy policy.

He is an elected director to the National Association of Environmental Professionals (NAEP) and received its national award for Outstanding Environmental Leadership.

Currently, he serves as an elected representative to the International Organization for Standardization's 242 working group, responsible for developing an ISO 50001 Energy Management System (EnMS) standard that will be used worldwide to manage the generation and use of energy. Eccleston developed and published the original concept, which has been adopted by a number of agencies around the world, for an integrated NEPA/ISO 14001 Environmental Management System.

He has served on two White House–sponsored Environmental initiatives. As part of this effort, he proposed, developed, and published the original concept for integrating the NEPA with an ISO 14001 Environmental Management System, which has now been adopted by a number of agencies around the world. He later generalized this approach to incorporate any international environmental impact assessment process. Still later, he generalized this integrated process to incorporate sustainable development.

xix

Eccleston is fluent on a wide range of environmental and energy policy issues such as sustainability, climate change, water and food scarcity, radioactive/hazardous waste, peak oil, population, and energy generation. His energy-related experience includes investigating nuclear, gas-fired, and coal-fired plants and renewable systems.

Frederic March is an environmental policy and planning consultant whose long career has encompassed energy and environmental systems for public infrastructure and private industry. He retired from Sandia National Laboratories where he was a principal analyst supporting its two ISO management system certifications, and was a project manager for Sandia's environmental, safety, and health programs. Prior to Sandia, he participated in many projects to develop energy and environmental policies and plans in the United States and in several other nations. He is currently a co-owner of Thinkwell Associates LLC, which offers a variety of training, policy and planning services. He has a B.S. degree in civil engineering from City University of New York, and a Master's degree in engineering systems analysis from the Massachusetts Institute of Technology. He is a coauthor of *Global Environmental Policy* and has published two prior books of compliance instruction pursuant to the National Environmental Policy Act.

Tim Cohen is a distinguished member of the technical staff at Sandia National Laboratories where he is currently responsible for leading the systems engineering efforts for Sandia's enterprise management system architecture as well as Sandia's ISO 9000 program. His previous responsibilities include strategic planning for science and technology, energy policy and planning analysis, and National Environmental Policy Act compliance for DoD and DoE facilities. He has over 20 years of experience leading agricultural and environmental resource analyses, environmental planning and impact studies, and economic impact studies of DoE facilities. He has B.S. and Master's degrees in agricultural economics from New Mexico State University. He also holds a Manager of Quality/Organizational Excellence certification from the American Society for Quality, as well as a Project Management Professional certification from the Project Management Institute.

Introduction

Product standards have existed for a long time, but the development of international standards for managing how an organization functions, rather than the nature of its product, is a modern invention that has been gaining popularity. The International Organization for Standardization (also referred to as ISO) is a body composed of representatives from various nations around the world whose mission is to develop common standards for products and services, including standards for management systems. It is widely recognized that to achieve the greatest improvement in either product quality or environmental performance, a management system like those defined by ISO must be in place.

History of the ISO 50001

U.S. involvement with an energy management system standard can be traced as far back as the year 2000, when the American National Standards Institute (ANSI) introduced its Management System for Energy (MSE 2005) standard. This was followed by two later versions.

In 2007, the United Nations Industrial Development Organization (UNIDO) hosted a meeting to study the concept of an energy management standard. The meeting led to a memorandum to the ISO Central Secretariat requesting that the ISO begin work on developing an international energy management standard. The reader is directed to Appendix A for an introduction to energy consumption and energy systems, while Appendix B describes how current trends and patterns of energy consumption establish the economic rationale for investments into significantly more effective and efficient energy systems to support our nation's industrial and public infrastructure services.

In parallel with the UNIDO initiative, U.S. interests approached the American National Standards Institute (ANSI) in 2007 to discuss how to promote an ISO Energy Management System (EnMS) resolution within the ISO. ANSI began working with these U.S. interests to develop an ISO EnMS proposal. The U.S. Department of Energy (DOE) also helped spearhead the effort to develop the new standard. ISO, in fact, has identified energy management as one of the top five fields meriting development of an international standard.

In 2008, the Technical Management Board of ISO approved a new project committee—ISO/PC 242—to develop an ISO 50001 energy management standard. The U.S. ANSI is serving as the co-secretariat for the committee. As with any ISO standard, there is ample opportunity for stakeholder input.

The first PC 242 meeting was held in September 2008 in Arlington, Virginia. Ninety participants from 25 countries, as well as UNIDO, attended. To ensure maximum compatibility with existing management systems, the key decision to be made at this meeting was whether to base standards on the common elements found in all of the ISO's management system standards (e.g., 9001, 14001). At the conclusion of this meeting, the participants had reached consensus on basing ISO 50001 on the foundation used in ISO 9001 and 14001. Because of the high priority given to this effort, the ISO decided to pursue a 2-year accelerated schedule under which an ISO 50001 EnMS standard would be published by 2011.

The ISO 50001 builds upon existing national standards and initiatives such as the U.S. ANSI MSE 2000:2008 and the European Union EN 16001:2009, and represents the latest and best thinking on the management of energy.

The purpose of the new standard

The purpose of the new standard is spelled out in the introduction to ISO 50001:

> The purpose of this International Standard is to enable organizations to establish the systems and processes necessary to improve energy performance, including energy efficiency, use, consumption, and intensity.
>
> Implementation of this standard should lead to reductions in energy cost, greenhouse gas emissions, and other environmental impacts, through systematic management of energy.

Thus, the impetus to reduce greenhouse gas emissions (GHG) and promote renewable and efficient use of energy sources provides a strong rationale for developing an EnMS standard. Experience has shown that most industrial energy efficiency efforts have been achieved through changes in how energy is managed, rather than through installation of new technologies. Some companies that have voluntarily adopted an energy management process (a central feature of an EnMS Standard) have

Introduction xxiii

reported major energy system improvements and savings. The new ISO standard provides organizations with management strategies to increase energy efficiency, reduce cost, and improve environmental performance, including reduced greenhouse gas emissions.

Corporations, suppliers, energy service companies, utilities, government agencies, and many other organizations are expected to embrace ISO 50001 as a critical tool for reducing energy usage and greenhouse emissions. The standard is designed to be general and flexible enough to be adopted by virtually any organization. The standard defines terms, establishes management system requirements, provides guidance for implementing the management standards, and requires metrics and measurements for assessing effectiveness.

The benefits to an organization include those shown in Table I.1, to name but a few.

The number of companies, organizations, and government agencies that will eventually adopt an EnMS around the world is easily in the tens of thousands. It has been widely estimated that this standard will affect 60% of the world's energy producers and consumers. The majority of these organizations will have no knowledge or expertise with implementing an EnMS. These organizations will have a strong need for resource guides (books) and consultants to assist them in designing and implementing their EnMS.

The potential and obvious savings in terms of reduced energy costs clearly point to the fact that the first major users of this standard will be industries. According to Piñero, Chairman of ISO PC 242, it is estimated that an EnMS will lead to potential long-term increases in energy efficiency of 20% or more for industrial facilities.

Relationships among ISO 9001, 14001, and 50001

ISO 50001 will complement the ISO 9001 and 14001 systems of standards for organizational and environmental management, respectively. The ISO 50001 standard is designed to manage energy across the entire international commercial sector, including industry plants and commercial facilities, as well as most other organizations. The standard applies to all factors affecting energy use that can be monitored and influenced by an organization. As explained in Chapter 1, ISO 50001 does not specify energy performance criteria. Instead it is designed to provide a general-purpose system that allows the user organization to choose performance standards and criteria that they deem best meet their requirements. The framework of the ISO EnMS is designed around the continual improvement plan–do–check–act approach utilized in ISO 9001 and ISO 14001. The EnMS is related to ISO 9001 and 14001 in the following manner:

Table I.1 Benefits of adopting an ISO 50001 energy management system

Improve efficient use of existing energy-consuming technologies and practices. Aid organizations in making better use of their existing energy-consuming technologies and practices. Present a framework for promoting energy efficiency throughout the supply chain

Reduce costs. Reduce energy costs via a structured approach for identifying, measuring, and managing energy consumption.

Prioritize adoption of new technologies and practices. Aid facilities in evaluating and prioritizing the incorporation of new energy-efficient technologies, including alternative energy systems and conservation measures.

Improve business performance. Drive greater productivity by identifying innovative technical solutions and affecting behavioral change to reduce energy consumption.

Promote environmental performance and reduce greenhouse gas (GHG) emissions. Reduce emissions and pollution, reduce resource consumption (e.g., consumption of fuels, water, etc.), and reduce greenhouse emissions.

Comply with regulatory requirements. Meet current or future mandatory energy efficiency targets and/or the requirements of GHG emission reduction legislation.

Engage top management. Position energy management in the boardroom as a key issue essential to sustaining a competitive business.

Formalize organizational energy policy and objectives. Provide a foundation for sound decisions, create respect for the energy management policy, and embed energy-efficiency thinking throughout an organization.

Integrate energy with other management systems. Align the EnMS with existing management systems for incremental benefit.

Secure energy supply. Understand the internal energy risk exposure and identify areas of the organization at greatest risk for short- and long-term changes in commercially available energy supplies and costs.

Drive innovation. Develop opportunities for new products and services in the low-carbon economy of the future.

Measuring, benchmarking, and reporting. Provide guidance on benchmarking, measuring, documenting, and reporting energy usage improvements, including projected impacts of GHG emission reductions.

Transparency and communication. Promote transparency and facilitate communication on generation/usage of energy.

- ISO 9001—provides a continuous improvement system for managing the quality of products and services provided by an organization.
- ISO 14001—provides a continuous improvement system for managing an organization's environmental compliance and reducing its environmental impacts.
- ISO 50001—provides a continuous improvement system for managing the production and organization's use of energy.

Introduction xxv

While the ISO 50001 standard has been designed to be used as an independent stand-alone system, it can be integrated with other management systems such as ISO 9001 and 14001, and other similar processes systems such as an occupational health and safety system.

ISO 50001 Energy management system process

The basic 7-stage EnMS process was described previously (see Figure 1). The basic components of each stage are described in more detail in subsequent chapters. As illustrated in this figure, an EnMS provides a structured system (i.e., "plan–do–check–revise") in which a set of management procedures is used to systematically identify, evaluate, manage, and address energy-related issues and requirements.

The ISO 50001 system is specifically designed to be used by any organization. An organization includes a firm, company, corporation, enterprise, authority, or institution, whether incorporated or not, or public or private, that has the authority to control its energy use and consumption; an organization can even include a person or group of people.

The organization should define the boundaries of the ISO 50001 management system. The term *boundaries* refers to the physical or site limits and/or organizational limits of the ISO 50001 system. Examples of boundaries include a particular process or group of processes, a plant, an entire site, or even multiple sites under the control of the organization. The term *energy* refers to the various forms of primary or secondary energy that can be purchased, treated, stored, recovered, or used in equipment or in a process (e.g., fuels, heat, steam, electricity, compressed air). The following sections provide a more detailed description of essential EnMS functions.

Table I.2 provides a brief overview of a typical ISO 50001 development and maintenance process. Subsequent chapters describe the process in more detail.

Improving EnMS system versus improving "energy performance"

It is important to note that the ISO standard draws a distinction between the terms *energy consumption* and *energy use*. Energy consumption refers to the quantity of energy that is actually consumed. In contrast, energy use denotes the manner or type of application of energy use including lighting, heating, cooling, ventilation, transportation, or production lines.

With respect to an ISO 50001, *energy performance* refers to measurable results related to energy use and energy consumption. This term tends to mean the act of reducing and enhancing energy efficiency and/or reducing energy-related environmental impacts, such that

Table I.2 Typical ISO development and maintenance process

Stage 1—Environmental Policy: The EnMS process is initiated with the preparation and establishment of an energy policy.

Stage 2—Planning: The next step in the EnMS process involves development of a plan for implementing the system. While the planning function is often performed to determine how an organization will meets its quality policy, it can also be used more comprehensively to develop detailed energy plans. Significant areas of energy use and consumption are identified, energy objectives and targets are established, and a program to achieve them is developed. This plan includes identification of

1. Energy Use and Consumption: Operations, activities, products, and services are reviewed to identify how they interact with and may affect the generation or usage of energy.
2. Objectives and Targets: Environmental objectives and targets are developed and communicated throughout the organization. A program is developed for achieving objectives and targets.
3. Legal and Other Requirements: The plan identifies legal and other energy-related requirements with which the organization must comply.

Stage 3—Implementation: Once the plan has been formalized, the EnMS is ready for actual integration and implementation with the organization's functions and activities. EnMS responsibilities are assigned. Employees are trained to ensure that they are aware of the plan and are able to perform required duties in compliance with the EnMS policy and plan. Specific work procedures are developed, defining how specific tasks are to be conducted. These implementation requirements are summarized below:

1 Structure and Responsibility:
   a. Roles, responsibilities, and authorities are defined for personnel whose activities may directly or indirectly affect the generation and/or use of energy.
   b. An individual(s) is appointed by top management as the "Management Representative(s)." The Management Representative(s) is assigned responsibility and authority for ensuring that the EnMS complies with the ISO 50001 standards and for reporting EnMS performance issues to top management.
2. Training, Awareness, and Competence:
   a. The organization identifies training requirements of personnel whose work may significantly impact the generation or use of energy. Personnel must receive appropriate education and training and/or have experience to deal with energy requirements.
   b. Communication: Communication of relevant information concerning energy policy, objectives, and targets, and EnMS is required throughout the organization.
   c. Energy Management System Documentation: Information must be maintained that describes the basics of the EnMS. The documents must be reviewed on a regular basis. This documentation must be managed and maintained through an established document control system (DCS).

Table I.2 Typical ISO development and maintenance process (Continued)

> d. Operational Control: Activities that can significantly impact energy use and consumption, and are relevant to the organization's objectives and targets, must be identified. The organization should ensure that these operations are performed according to the EnMS plan to ensure they are performed under *controlled conditions*. Controlled conditions can include documented procedures with specific operating criteria.
>
> **Stage 4—Monitoring and Corrective Action**: This stage involves checking and audits, control of nonconformances, corrective action, and preventive action. Characteristics of operations and activities that can significantly impact energy generation/usage need to be regularly monitored and measured. Monitoring and measurement results need to be compared with legal and other requirements to assess compliance.
>
> **Stage 5—Management Review**: The final stage involves a review by the organization's management of the EnMS. This step helps ensure that the system is operating effectively and provides the opportunity to address changes that may be made to the EnMS.

the quality of energy generation/usage is improved. Instead of focusing on energy performance, the ISO 50001 focuses on improving the *management process* that manages or administers an organization's functions and activities involving energy (Table I.2). This International Standard considers all types of energy; the term *energy* includes renewable, non-renewable, and recovered energy.

Thus, ISO 50001 does not actually require that an EnMS improve energy or environmental performance (i.e., enhance energy efficiency or reduce environmental impacts). For instance, ISO 50001 does not prescribe a particular level of energy performance that an organization must meet, require use of particular energy technologies, or establish regulatory standards for energy outcomes. In fact, some organizations engaged in similar activities may have widely different effects on the energy generation/usage, including environment impact, and yet all comply with the ISO 50001 standard.

The focus of an ISO 50001 EnMS is on improving management processes, practices, and procedures that control an organization's functions and activities, which can affect energy performance as well as environment impacts. The overarching intent is that by implementing a *management process* that administers an organization's functions, products, and services, and by continually improving this management system, this process will eventually lead to improved energy performance.

Although this is generally true of organizations that are truly committed to the goal of improving environmental quality, it may not be true of an organization that lacks a serious commitment; in this case an EnMS may amount to nothing more than "window dressing" to improve a business' or organization's image with the public and consumers.

It is important to note that adherence to the ISO 50001 standard does not, by itself, release an organization from full compliance with other local or national environmental laws and regulations regarding specific environmental performance standards that must be met. In fact, it provides procedures to help ensure that all applicable laws and regulations have been identified, as well as an auditing/monitoring procedure to identify any noncompliances.

The value of an EnMS in addressing peak oil

When Dr. King Hubbert, a world-famous geophysicist, stepped up to the stage in 1956, he probably did not completely understand the effect that his ideas would have on our future. In 1954, Hubbert presented a paper stating that petroleum production would follow a typical statistical "bell-shaped curve." He noted that the quantity of oil available for production in any given region must necessarily be finite, and therefore subject to depletion at some point.

Whenever a new oil field is discovered, the petroleum yield from that location tends to increase rapidly for a period of years, as drilling infrastructure is put in place and extraction activities are ramped up. Once approximately half of the oil field's reserves are pumped out, however, the oil source reaches its peak rate of production. Then decline sets in, with the rate of production decrease ultimately approximating a "mirror image" of the production increase rate seen in the oil field's early years. Using his high estimate for U.S. oil reserves, Hubbert predicted that U.S. oil production would peak in 1970. Many scientists dismissed Hubbert.

In the years following Hubbert's paper, U.S. oil production continued its steady upward progression, and 1970 came and went. But trends are not always immediately apparent. It can take a few years of data before they become clear. And soon enough, the statistics were in and they were indisputable: Hubbert had been dead on. U.S. oil production had in fact peaked in 1970—exactly the year he had predicted based on his "high oil inventory" estimate. For a more detailed review of this pressing issue, the reader is directed to Appendix C.

World peak?

Hubbert's prediction went much further. He argued that his theory applied not only to the U.S. petroleum industry, but to global oil production as well. Throughout the 1960s and 1970s, Hubbert continued to refine his theory. He eventually predicted that a peak, also referred to as "Peak Oil," in world oil production would occur around 2000.

Introduction *xxix*

Doe's peak oil assessment

Given the chilling implications of Peak Oil, a study (the Hirsch Report) of the Peak Oil question commissioned by the U.S. Department of Energy paints a sobering picture of the problem—and the level of effort needed to address it.

The Hirsch report states that peak oil is a "unique challenge," something the world has never before faced. The report's executive summary opened with an ominous sentence:

> The peaking of world oil production presents the U.S. and the world with an unprecedented risk management problem.

Among the key conclusions outlined in the report were the following:

- World oil production will peak, although experts differ on exactly when the peak will arrive.
- Peak oil will have a severe impact on the economy.

The authors note:

> Without mitigation, the peaking of world oil production will almost certainly cause major economic upheaval. However, given enough lead-time, the problems are solvable with existing technologies.

The authors go on to note that the "obvious conclusion" from their overall analysis "is that with adequate, timely mitigation, the costs of peaking can be minimized. If mitigation were to be too little, too late, world supply/demand balance will be achieved through massive demand destruction (shortages), which would translate to significant economic hardship."

There is some evidence that global oil production may already be in the process of peaking and that our future may be characterized by disruptions and soaring energy prices. Even today, few policymakers appear to fully appreciate the consequences that peaking could have on the entire global food chain and industrialization, which are currently dependent on cheap fuel. Following the peak, there could be ominous worldwide food shortages; some go so far as to warn that as oil production declines, so must the human population. The Western way of life will be in particular danger because it has built its entire infrastructure and society around relatively abundant, stable, and cheap oil. It is becoming increasingly apparent that the era of "cheap" oil is coming to an end. All of this points to the need for sustainable energy systems (see Appendix D) and widespread adoption of EnMSs to help manage existing energy usage.

Concluding thoughts

The implications of a peak oil scenario on our modern industrial society are almost inconceivable. Consider for a moment one single segment—agriculture. In his book, *The End of Food*, Paul Roberts reports that malnutrition was common throughout the 19th century. It was not until the 20th century that cheap oil allowed agricultural output sufficient to avert famine. It has been widely argued that an exponential increase in energy supply is the principle reason that we have produced a food supply that has grown exponentially in parallel with the increase in human population. Thus, we have avoided wide-scale famine largely because fossil fuel supplies expanded geometrically. Nearly 20% of all energy used in the United States is funneled into our agricultural system. If this energy supply begins to tighten and prices escalate, the world could face a nightmare scenario.

The multifaceted failure of a substantial portion of modern industrial civilization is so under-appreciated or completely misunderstood by most policymakers that we are virtually unprepared to deal with the outcome. The failure to inject the reality and potential impacts of peak oil into the mainstream public policy forum is a grave threat to our modern society.

An overview of greenhouse gas emissions and potential climate change is presented in Appendix E. The issues of increasing greenhouse emissions and peak oil are just two of the key reasons that creation of an ISO 50001 standard to manage the generation and use of energy is an international priority.

chapter one

General requirements for an energy management system

This chapter describes the general requirements that an organization must comply with to qualify for ISO 50001 registration.

1.1 Qualifications

> (a) In order to qualify for ISO 50001 registration, an organization is required to establish, document, implement; and maintain an energy management system (EnMS) in accordance with the requirements of this international standard.

1.1.1 What does this mean?

An energy system consists of a set of processes that act together to serve a given function such as indoor climate control, product realization, and transportation of materials and people. An energy management system establishes performance goals, objectives, and targets for that system, monitors performance, and implements preventive and corrective actions.

1.1.2 What are the benefits?

A systematic and functional energy system meets or exceeds the goals, objectives, and targets established by management. Meeting the ISO 50001 standard will foster an organizational culture of continuous improvement in energy efficiency, provide reliable documentation of metrics that monitor actual performance, help management make effective decisions, enhance public relations, and perhaps above all, improve the corporate bottom line (see the section on benefits in the Introduction for more detail).

1.1.3 How do we implement this?

The ISO 50001 standard provides a high-level systematic guide for implementing a new energy system or improving the performance of an existing one. This book augments the ISO 50001 requirements with more detailed practical guidance on how to meet those requirements. Implementation of the EnMS flows by way of the following steps and is described in detail in corresponding chapters:

1. Developing an energy policy (Chapter 3 of this book)
2. Performing an energy planning process (Chapter 4 of this book)
3. Physical implementation and operation of the EnMS (Chapter 5 of this book)
4. Checking performance (Chapter 6 of this book)
5. Management review of that performance (Chapter 7 of this book)
6. Assessing the effectiveness of the energy policy in driving the desired results

Also see Figure I.1 in the Introduction to this book.

1.1.4 What evidence is needed?

The evidence needed includes documentation of the policy, plan, and performance of the system as further defined in subsequent chapters.

1.2 Requirements for organizations

> (b) Organizations are required to define and document the scope and boundaries of the EnMS in accordance with the requirements of this international standard.

1.2.1 What does this mean?

The "scope" means the specific organizational functions with processes that produce or consume energy, and that management defines as being within the scope of its EnMS. "Boundaries" refer to the physical limits of energy flows that are managed by the system. For example, the meter that measures electric energy input from a supplier and the stacks that release spent energy and pollutants to the atmosphere may each constitute a system boundary. In addition, there may be specific operations or functions that need to be isolated from the organization's overall energy management system.

Chapter one: General requirements for an energy management system 3

1.2.2 What are the benefits?

Defining the scope and boundaries enables the organization to qualify for ISO registration without allocating scarce resources to functions that are not considered necessary in managing the organization's energy use.

1.2.3 How do we implement this?

Performing a review of energy flow and processes will enable management and staff to decide what falls within or outside the scope and boundaries of the EnMS.

1.2.4 What evidence is needed?

A documented statement should define and justify the scope and boundaries.

1.3 EnMS requirements

> (c) And finally, organizations are required *to* determine and document how the EnMS will meet the requirements of the ISO Standard so as to achieve continual improvement of its energy performance and its EnMS.

1.3.1 What does this mean?

It means the full set of documentation requirements described in Chapters 2 through 7. To qualify for registration, one must demonstrate to an ISO auditor that the EnMS meets the requirements of the ISO standard. It also means that effective performance to satisfy or exceed management expectations can only be achieved through a deep-seated commitment to continual improvement. Those in charge of the EnMS need to provide substantive evidence as to why further improvement is not practical or feasible.

1.3.2 What are the benefits?

The major benefit should be substantial reduction in consumption of energy as well as a cost savings in terms of operating expenses.

1.3.3 How do we implement this?

Implementation requirements of continual improvement are provided in Chapter 6. Successful implementation requires more than following a script of requirements. It requires a management attitude that encourages the entire team to

- Think outside the box.
- Be unafraid of questioning existing practices.
- Not seek perfection, but simply take the initial effort to begin and continue to improve.
- Think of reasons for doing it, rather than excuses for why it cannot be done.
- Search for root causes that are responsible for problems.
- Discard conventional ideas that will not solve the problem at hand.

1.3.4 What evidence is needed?

The cumulative and comprehensive evidence needed to support ISO 50001 registration is provided in Chapters 2–7.

chapter two

Management responsibility

This chapter describes the organization's management responsibilities (ISO 50001, Section 4.2). We begin by describing the requirement for top management to demonstrate its commitment and support to the implementation of an ISO 50001 Energy Management System (EnMS).

2.1 General requirements (4.2.1 of the ISO standard)

The ISO 50001 standard requires top management to demonstrate its commitment and support to the EnMS and to continually improve its effectiveness.

2.1.1 What does this mean?

The actions undertaken by top management to support the energy management system must be communicated and made visible to the rest of the organization. This also implies that top management fully understands what actions they must take to support the EnMS as it is developed. Increasing an organization's energy performance requires the participation of every principal part of the organization. However, only the persons or people who direct and control the organization at the highest levels have the authority, influence, and control of resources necessary to sustain and improve an EnMS. Although they will not do the majority of the work required, it is they who must inspire employees to change their behavior.

Improving energy performance begins with a personal commitment by top management. However, for this commitment to ignite organization-wide participation, it must somehow be transferred to the rest of the organization. Top management should realistically expect their employees to adopt only those energy performance behaviors that they themselves are willing to adopt.

2.1.2 How will this benefit us?

Top management is responsible for ensuring the consistent and visible involvement of management in developing and implementing the EnMS. Communicating with the organization and demonstrating commitment by exhibiting visible behaviors will engage the rest of the organization and make the organization's energy policy words to live by rather than words hanging on a wall.

2.1.3 How do we implement this?

First, identify who the top management is in the organization. In a small business, it could be the owner or the delegated office manager. In a large corporation, it could be the CEO and the vice-presidents. If the scope and boundary of the EnMS is a business unit that is part of a larger company, the top management could be the business unit vice-president or a director. Whatever frame of reference is being considered, identify the people who direct and control the organization at the highest levels and have the authority, influence, and control of resources.

Specifically, top management demonstrates its commitment by implementing the actions described in Sections 4.2.1a–j of the ISO standard. Note that top management does not have to do everything themselves. They can demonstrate their commitment by taking specific actions themselves, and by delegating actions to others. Providing resources, making decisions, assigning job responsibilities, and communicating are examples of actions personally undertaken by top management. Dedicating the needed time to understand the EnMS in the context of the overall business is an often underappreciated demonstration of management commitment. When top management uses their most limiting resource, their time, to consider their energy performance, they are most assuredly making a commitment. This understanding allows them to ensure that the right processes and procedures are in place, aids them in prioritizing energy-related expenditures, and helps them do a better job of communicating the organization's energy performance core values. Sometimes, top management may have trouble becoming passionate about energy performance even though they support it with financial resources. That's okay! Expending their second most limiting resource, money, on maintaining and improving the EnMS supported by timely decision-making is an extremely effective form of commitment.

In addition to personally taking actions, top management also demonstrates their commitment by ensuring that specific actions are taken by someone else in the organization. The sections below define what actions top management must perform themselves and those which they may delegate to others.

2.1.4 What evidence do we need?

Evidence of top management commitment comes in several different forms, including documentation, communications, the existence

Chapter two: *Management responsibility*

of energy performance-related actions that have been taken, and organizational understanding. Much of the evidence used by top management as well as ISO 50001 auditors to review the adequacy and performance of the EnMS will be in the form of hardcopy and electronic documents supporting the actions described in Sections 4.2.1a–j of the ISO standard. As an example, the organization's energy policy is often displayed in a public area. This documented evidence could be supplemented by testing whether employees are aware of, understand, and know how their actions personally support the energy policy. Gauging the level of top management commitment requires a holistic view of the organization and its energy performance activities.

2.2 Establish, implement, and maintain the energy policy (4.2.1a of the ISO standard)

> Top management is required to establish, implement, and maintain the energy policy.

2.2.1 What does this mean?

This means that top management must formally express the organization's overall intentions and direction related to its energy performance. These intentions must be documented, communicated, and understood by everyone in the organization. The energy policy typically contains high-level statements such as, "We will continually improve our energy performance" or "We will meet all customer and applicable legal energy-related requirements." The energy policy itself is described in detail in Chapter 3.

2.2.2 How will this benefit us?

Establishing an energy policy ensures that the organization's energy needs are understood and provides direction to the entire organization, leading to visible and expected results. Implementing the energy policy provides the framework for action and for the setting of energy objectives and energy targets. See Chapter 4 for more detailed information about setting energy objectives and targets.

2.2.3 How do we implement this?

Top management appoints an "energy representative" and approves an energy management team (see Section 4.2.1b of the ISO standard) to draft a policy for top management adoption and continues to play its role as defined in Chapter 3.

2.2.4 What evidence do we need?

A documented and approved statement of top management's overall intentions and direction for the organization related to its energy performance is needed. It must state the organization's commitment to achieving continual improvement in energy performance, meeting the energy requirements to which it subscribes, and ensuring the necessary data and resources will be made available for doing so. A more detailed description of the energy policy requirements is provided in Chapter 3.

2.3 Management representative and energy management team (4.2.1b of the ISO standard)

> Top management is required to appoint a management representative and approve the formation of an energy management team.

2.3.1 What does this mean?

This means that the management representative is responsible to top management and is fully authorized to manage all aspects of the energy system as it evolves. Top management is responsible for approving the formation of an energy management team because they are the ones who control job assignments and funding. The energy management team is responsible for ensuring the implementation of energy management decisions take action. The size of the team should be commensurate with the size and complexity of the organization. In some instances particularly in the of small organizations, the management representative could also be the energy management team. Keep in mind that members of the energy management team need not work full time improving energy performance and may also have other responsibilities besides improving energy performance.

2.3.2 How will this benefit us?

Appointing a management representative and forming an energy management team benefits the organization by providing the formal structure, resources, and actions by which energy performance can be improved. Top management can rely on a single person for accurate information on the EnMS and changes in energy performance that enables them to make informed and timely decisions. The formal appointment of the management representative provides the management backing required to work throughout the organization to establish, maintain, and improve the EnMS. A formal energy management team gives visibility and recognition

Chapter two: Management responsibility 9

to the organization's energy performance efforts. Empowering employees to develop and implement energy performance improvements will create beneficial and lasting change throughout the organization.

2.3.3 How do we implement this?

Energy performance does not just improve all by itself. Top management assigns a responsible individual, who in turn forms an energy management team. The management representative along with the energy management team, all supported by top management, engage the entire organization and improve the EnMS to obtain improved energy performance. The responsible management representative should have the appropriate skills and competence as well as the authority to:

- Ensure the organization's EnMS is established, implemented, maintained, and continually improved in accordance with the ISO 50001 standard
- Report to top management on the performance of the EnMS and changes in energy performance
- Form the energy management team
- Plan and direct energy management activities designed to support the organization's energy policy
- Define and communicate responsibilities and authorities to facilitate effective energy management
- Determine criteria and methods needed to ensure effective operation and control of the EnMS

Additional guidance pertaining to management representatives is provided in the roles, responsibility, and authority section a little later.

2.3.4 What evidence do we need?

The organization should document the name or specific positional title of the energy management representative. This could be accomplished in management review meeting minutes or wherever the organization's roles, responsibilities, and authorities are documented. The energy management representative should document the names of the energy management team members.

2.4 Provide energy management system resources (4.2.1c of the ISO standard)

> Top management is required to provide the resources necessary to establish, implement, maintain, and improve the EnMS.

2.4.1 What does this mean?

Establishing an EnMS requires an investment of employees' time, competency, training, funding, and facilities. For example, some employees will use organizational resources to attend training, conduct audits, or implement actions designed to meet energy objectives and targets. Financial resources may be needed to maintain and purchase capital equipment in order to increase energy performance. Top management, as their name implies, are the ones who control all these resources and can authorize their use.

2.4.2 How will this benefit us?

This provides the organization with a better understanding of the relationship between energy costs and improvement benefits. Increased energy performance from implementing the actions described in this book is enabled by the appropriate levels and types of resource commitments.

2.4.3 How do we implement this?

This is implemented through management decisions and subsequent allocations of resources to support the management representative and energy management team efforts to increase energy performance. These decisions should be supported by any needed benefit-cost analyses, and financial and business considerations. The management review is an appropriate forum for discussing and making these types of resource allocation decisions.

2.4.4 What evidence do we need?

Your organization will need records documenting funding allocation decisions and budget allocations as well as for satisfying an ISO 50001 auditor. The auditor would also observe the activities of the management representative and the energy management team to infer that resources are available to increase energy performance.

2.5 Defining the energy management system scope and boundaries

> Top management is required to identify the boundaries and scope addressed by the EnMS.

2.5.1 What does this mean?

The ISO 50001 EnMS requirements need not apply to the entire organization. Each organization has the flexibility to define their own scope and boundaries for themselves. Boundaries can be based on geographic, physical, organizational, or other criteria. For example, a multinational corporation might define their EnMS to include only their North American facilities. A company that manufactures several different products might include only a single product within the bounds of their EnMS.

2.5.2 How will this benefit us?

Bounding the EnMS and defining the scope help the organization to focus its operational efforts where the net savings from energy investments can really make a difference while avoiding wasted effort that will not produce a "significant" energy saving. This helps create a balance between the organization's business priorities and its energy performance commitments.

Defining the scope and boundaries of the organization's energy management system helps the organization identify the energy-related international, national, regional, and local requirements with which the organization must comply. The organization's energy review will be much easier to develop because the scope and boundaries help clarify the precise energy sources, use, and consumption that will be measured, monitored, and improved.

Some organizations may find it difficult to establish an EnMS throughout their entire organization. The flexibility to establish the specific scope and boundaries provides the opportunity to pilot the EnMS in a subset of the organization and then propagate across the entire organization when it is sufficiently mature. This benefits the organization by defining precisely which employees are responsible for complying with the EnMS.

2.5.3 How do we implement this?

The members of management who are considering adopting an EnMS should consider what part of their organization is considered "top" management. A vice-president responsible for a single business unit within a multiproduct organization would be considered "top management" for their business unit because he controls people's time, job assignments, physical resources, and funding. However, he may have limited authority or influence within the entire organization as a whole. In this situation, the VP should initially scope the EnMS to include only his business unit. In this situation, the "organization" would be defined as the business unit. Furthermore, if the business unit was responsible for several plants, the VP might scope the EnMS to include only two of the plants if he wished.

2.5.4 What evidence do we need?

The scope and boundaries of the organization's EnMS should be documented and controlled according to the organization's procedures for controlling documents. For purposes of ISO 9001 registration, an auditor will speak with various employees to determine whether they understand if and how the EnMS applies to them and their work.

2.6 Communicating the importance of energy management

> Top management is required to communicate the importance of energy management to their organization.

2.6.1 What does this mean?

Top management should make their commitment to and the importance of energy management known throughout the organization. This includes communicating the energy policy, responsibilities and authorities, and the operational controls needed to improve energy performance.

2.6.2 How will this benefit us?

Communicating the importance of energy management helps build awareness and understanding throughout the organization, which will lead to commitment and actions for improving energy performance. It is one of the key elements for successful implementation of the energy program.

2.6.3 How do we implement this?

This is accomplished by top management expressing its thoughts and feelings, or giving information to the employees throughout the organization. This can be implemented through writing, speaking, or actions using the organization's communication channels.

- Examples of communication through writing include memos, e-mails, newsletters, flyers, internal web teasers, etc.

- Examples of verbal communication that your organization may be using include staff meeting briefings, town hall meetings, one-on-one conversations, new hire orientations, voice-mails, etc.
- Examples of demonstrating visible action on the part of top management might include adhering to the EnMS, taking action to improve energy performance, and sponsoring or participating in energy conservation awareness events.

2.6.4 What evidence do we need?

For purposes of ISO 50001 registration, auditors may typically look for documented examples of e-mails, flyers, meeting minutes, memos or any other documents and records demonstrating that top management has communicated the importance of energy management throughout the organization. Auditors may also speak with employees to gauge their level of awareness and understanding as a means of assessing the effectiveness of the communication.

2.7 Establishing energy performance objectives and targets (4.2.1f of the ISO standard)

Top management is required to ensure that energy performance objectives and targets are established for the organization.

2.7.1 What does this mean?

As part of the energy planning process, *energy objectives* and *targets* are established to meet the goals documented in the organization's *energy policy*. Top management must ensure that this is done as part of the energy planning process. To determine whether these metrics are met, top management should ensure that performance requirements for the energy objectives are measurable. More detailed information is provided in Section 4.6 of Chapter 4.

2.7.2 How will this benefit us?

The energy objectives define the specific achievements that would enable the organization to meet its energy policy and improve energy performance. They help the organization identify the specific actions needed to define operational controls and improve energy performance. Energy targets help the organization to quantifiably determine if they are meeting their energy objectives.

2.7.3 How do we implement this?

Energy objectives and targets are initially established as part of the organization's energy planning activities. They are derived from and consistent with the energy policy and give rise to the organization's energy action plan for achieving them. A more comprehensive explanation of how to establish, implement, and maintain energy objectives and targets is provided in Section 4.6 of Chapter 4.

2.7.4 What evidence do we need?

The organization should maintain and control a documented record of its current energy objectives and targets. Management review records are useful as evidence that top management has reviewed and approved the organization's energy objectives and targets. More detailed information is provided in Section 4.6 of Chapter 4.

An *energy objective* is a specified outcome or achievement established to meet the organization's policy to improve energy performance. In contrast, an *energy target* is a detailed and measurable (e.g., quantifiable) energy performance requirement that is established for achieving the energy objective. In other words, "objectives" refers to general long-term goals such as increased energy efficiency, cost reductions, and development of better employee energy practices, training, or improved energy-related communication with other interested parties.

In contrast, a "target" refers to more specific, measurable events such as the reduction of energy utilization by 20% or a reduction in CO_2 emissions by 10%. Table 2.1 shows the difference between objectives and targets, as well as assignment of responsibilities for ensuring that these objectives and targets are met.

While the energy objectives tend to apply across the entire organization, targets often vary over time and across various organizational functions and activities. In establishing and reviewing its objectives and targets, the organization considers:

- Legal and other energy-related requirements
- Significant energy uses
- Opportunities for improving energy performance as identified in the energy review

The organization also needs to consider its

- Financial, operational, and business conditions
- Technological options
- The views of interested parties

Table 2.1 Example of Energy Objectives and Targets (for Illustrative Purposes only)

Objectives	Targets	Responsibility
1. Energy compliance: Comply with all applicable energy-related environmental laws and regulations	Zero penalties or fines per year	Principal regulatory manager
2. Minimize wasted energy	To reduce energy expenditures, re-cycle 75% of all paper products, and 50% of aluminum waste	Chief process engineer
3. Conserve energy	Reduce electricity consumption by 20%	Chief plant engineer
4. Improve the ENMS	Obtain ISO 50001 certification	ENMS program manager

The organization develops and implements energy management *action plans* for achieving its objectives and targets. The energy management action plans include:

- Designation of responsibility for achieving the objectives/targets
- Means and time frame in which individual targets will be achieved
- Statement of the method by which energy performance improvements are verified
- Statement of the method that will be used in verifying the results of the action plan

The energy management action plans are documented and updated at regular intervals. Objectives and targets should be defined for appropriate functions and levels of the organization. Work procedures, instructions, and controls should be developed to ensure implementation of the policy and that the targets can be achieved.

2.8 *Energy performance indicators (4.2.1g of the ISO standard)*

> Top management is required to ensure that the energy performance indicators are appropriate to the organization.

2.8.1 *What does this mean?*

It is up to each individual organization to define how it will quantify and measure its energy performance. The term "Appropriate" means that the energy performance indicators are suitable and tailored for a particular organization. When it comes to energy performance indicators, there is no one-size-fits-all policy. Just as no two people are exactly alike, the same applies to organizations. Organizations differ in size, complexity, location, health, location, autonomy, energy use, etc. An organization's energy performance indicators should be custom-designed to conform to its own energy performance objectives. More detailed information is provided in Section 4.5 of Chapter 4.

2.8.2 How will this benefit us?

Energy performance indicators help the organization achieve energy performance improvements and meet other performance criteria. Analyzing changes in energy performance allows the organization to determine which actions are improving energy performance and to identify additional improvement opportunities.

2.8.3 How do we implement this?

Energy performance indicators are identified as part of the organization's energy planning activities to determine how one will monitor and measure energy performance. Choose indicators that reflect the organization's actual energy use. Once these indicators are defined, develop a quantitative method for how they will be calculated and updated. An energy performance indicator could be a simple metric ratio (e.g., energy consumed divided by the appropriate energy target) to a very complex model. More detailed guidance on implementing energy performance indicators is provided in Section 4.5 of Chapter 4.

2.8.4 What evidence do we need?

Evidence that energy performance indicators are appropriate to the organization can be found in document records of energy performance planning, review, and assessment. Additional evidence may be found by observing whether the organization is making progress in improving its energy performance. Organizations that use indicators not tailored to their energy use will find that their energy objectives and targets do not impact their energy performance. Energy action plans will be of no consequence because the organization will never know what is working and what is not. More detailed information is provided in Section 4.5 of Chapter 4.

2.9 Including energy considerations in long-term planning (4.2.1h of the ISO standard)

> Top management is required to include energy considerations in long-term planning, if applicable.

2.9.1 What does this mean?

Long-term business planning describes the process organizations use to define their future directions, develop strategies, and decide how to allocate resources accordingly. For those organizations that conduct long-term planning, some basic energy planning should be integrated with the overall long-term business planning. Energy should not be a stand-alone initiative considered after the fact.

2.9.2 How will this benefit us?

Organizations benefit because energy performance improvement becomes part of the culture, leading to better results. Energy-related capital equipment investments have a better chance of being funded if they are considered early when funding decisions are being made.

2.9.3 How do we implement this?

Include the organization's sources of energy, energy performance, and energy performance improvement in strategic planning sessions. Establish energy-related budget line items so that funding for the EnMS is made available. Identify any needed future energy-related investments so that they may be prioritized along with other investments at the appropriate time.

2.9.4 What evidence do we need?

The organization's strategic plan, business plan, or any other planning documents should demonstrate that energy has been considered as part of the overall long-term planning effort. Organizational budgets containing energy-related line items and the existence of energy-related capital equipment also may serve as indirect evidence that energy considerations have been included in long-range planning.

2.10 Ensure that results are measured and reported (4.2.1i of the ISO standard)

> Top management is required to ensure that results are measured and reported.

2.10.1 What does this mean?

Systematic and continuous improvement in energy performance requires the organization to know their current energy performance, their desired energy performance, and how to achieve the desired energy performance. The most proactive approach to energy performance improvement is to measure your energy performance and report it so that improvement actions can be identified if necessary.

2.10.2 How will this benefit us?

The organization benefits because it can proactively control energy performance based on information gained from measuring results. Top management is then able to make decisions based on fact rather than conjecture.

2.10.3 How do we implement this?

This requirement is initially implemented by top management communicating their expectations of reviewing quantitative energy performance measures as part of their management reviews. Further implementation involves establishing energy performance indicators and periodically checking the energy processes and products against the energy policy and objectives. These measures, along with the appropriate analysis, and preliminary conclusions and recommendations, are then reported to top management for use in their management review. Additional guidance on how to define effective measures and measure energy results is presented in Chapter 6.

2.10.4 What evidence do we need?

Evidence of measuring results includes documented plans and reports containing performance data, measures, and analysis. Evidence of reporting includes documented management review meeting agendas, reviews, and minutes. More detailed information is provided in Chapter 6.

2.11 Conduct management reviews (4.2.1j of the ISO standard)

> Top management is required to conduct management reviews.

2.11.1 What does this mean?

At planned intervals, top management should personally review the organization's EnMS and its performance to ensure its continuing suitability, adequacy, and effectiveness. More detailed information is provided in Chapter 7.

2.11.2 How will this benefit us?

The organization benefits from management reviews because top management becomes knowledgeable about the EnMS performance and can control future performance by identifying opportunities to improve and take appropriate actions.

2.11.3 How do we implement this?

Top management must personally conduct the management reviews. The organization should define how often the reviews will occur. Detailed guidance on management reviews is provided in Chapter 7.

2.11.4 What evidence do we need?

Agendas, information used during the review, and meeting minutes are all useful. Meeting minutes are especially useful for documenting decisions and action items and accessing them at a later time. More detailed information is provided in Chapter 7.

2.12 Roles, responsibility, and authority (4.2.2 of the ISO standard)

2.12.1 Appoint a competent management representative

Top management must appoint a competent management representative who has the responsibility and authority to perform a defined set of energy management activities.

2.12.1.1 What does this mean?

Top management must appoint an energy management representative to serve as their eyes and ears in energy-related matters and act in the capacity of an energy program manager. The energy management representative, who is usually a member of management, should be vested by top management with the responsibility for helping the organization achieve its energy objectives and targets and for improving energy performance. The management representative should have the appropriate knowledge, skills, and abilities needed to perform the required duties and be capable of exerting influence throughout the organization to implement and improve the EnMS. It is not necessary that the role of energy management representative be a full-time job. Anyone meeting the appropriate criteria may serve in this capacity, irrespective of their other organizational responsibilities. This should include the authority to direct financial and human resources to that end.

2.12.1.2   How will this benefit us?

Appointing a management representative and forming an energy management team benefits the organization by providing the formal structure, resources, and actions by which energy performance can be improved. It provides a consistent and comprehensive approach and clarifies roles and responsibilities, and linkages to all interested parties. Top management can rely on a single person for accurate information on the EnMS and changes in energy performance, which enables them to make informed and timely decisions. The formal appointment of the management representative provides the management backing required to work throughout the organization to establish, maintain, and improve the EnMS. A formal energy management team gives visibility and recognition to the organization's energy performance efforts. Empowering employees to develop and implement energy performance improvements will create lasting change throughout the organization.

2.12.1.3   How do we implement this?

Top management uses the aforementioned criteria to identify a single person to be their energy management representative. The appropriate responsibilities and authorities are documented wherever the organization documents its roles, responsibilities, accountabilities, and authorities. Sections 4.2.2a–g of the ISO standard define the activities essential for implementing and successful EnMS.

2.12.1.4   What evidence do we need?

An approved energy management charter identifies the energy management representative, the energy management team, roles and responsibilities, and the management reporting mechanism.

2.13   Establish, implement, maintain, and continually improve the energy management system (4.2.2a of the ISO standard)

The energy management representative is required to ensure that an EnMS is established, implemented, maintained, and continually improved in accordance with the ISO 50001 standard.

2.13.1   What does this mean?

While top management is ultimately responsible for the organization's EnMS, they typically delegate the authority for establishing, maintaining, implementing, and continually improving the energy processes and procedures to the energy management representative. This includes all the elements of the ISO 50001 standard that combine to ensure a successful EnMS. Everyone in the organization interacts with the EnMS in some fashion; however, the management representative is responsible for ensuring that all needed energy processes and procedures are established and designed to work together as a system.

An implemented EnMS means that the organization's employees perform the required activities and adhere to its requirements. Maintaining the EnMS requires taking appropriate actions to sustain its existence and ensuring it is not subject to the whims of every change in top management. Improving the EnMS means that the energy processes and procedures are revised periodically to become more useful and effective, leading to improved energy performance.

2.13.2 How will this benefit us?

Increased energy performance results from an EnMS that has a dedicated program manager with a direct line to top management.

2.13.3 How do we implement this?

- Form an energy management team
- Identify the needed energy processes and procedures
- Develop them or integrate the energy-specific requirements into existing quality, environment, or safety processes, and procedures
- Communicate expectations throughout the organization
- Implement the processes and procedures
- Measure, monitor, and improve as necessary

2.13.4 What evidence do we need?

- Documented energy processes and procedures
- Documented records of implementation
- Documented plans and reports

2.14 Report to top management on the performance of the energy management system (4.2.2b of the ISO standard)

> The energy management representative is required to report to top management on the performance of the EnMS.

2.14.1 What does this mean?

The management representative communicates (verbally and in writing) with top management about how their EnMS is performing. The management representative should work with top management to understand their expectations as to how often the communication will take place, how much time is to be allotted, what topics are to be discussed, and the medium for communication.

2.14.2 How will this benefit us?

An organization whose top management understands their EnMS performance benefits from informed management actions and decisions aimed at improving energy performance.

2.14.3 How do we implement this?

A well-constructed management energy review that meets the requirements of this standard will include information on the performance of the organization's EnMS. The management representative is responsible for providing that information in the form of verbal communication, written reports, viewgraphs, or any other means appropriate to the organization. The information should include the following at a minimum: context, information, analysis, assessment, and preliminary recommendations. Top management should not have to spend an inordinate amount of their review time to understand and determine the meaning of energy data.

2.14.4 What evidence do we need?

Relevant evidence includes EnMS performance reports, meeting agendas, meeting minutes, and any other records indicating communication has taken place.

2.15 Report to top management on changes in energy performance (4.2.2c of the ISO standard)

> The management representative is required to report to top management on changes in energy performance.

2.15.1 What does this mean?

When the organization's energy performance changes for the better or worse, top management needs to know about it. And it is the management representative's job to tell them.

Chapter two: Management responsibility 23

2.15.2   *How will this benefit us?*

Improved energy performance results from timely corrective actions when performance is not going as expected.

2.15.3   *How do we implement this?*

This could be included in the regularly scheduled briefings that the management representative provides to top management during management review meetings or other communications.

2.15.4   *What evidence do we need?*

Relevant evidence includes EnMS performance reports, meeting agendas, meeting minutes, and any other records indicating communication has taken place.

2.16   *Forming the energy management team (4.2.2d of the ISO standard)*

The management representative is required to identify authorized people to support energy management activities as appropriate.

2.16.1   *What does this mean?*

While it is possible and allowable by the ISO 50001 standard for the management representative to do all the energy-related work for the organization, it is highly unlikely that this will be the case. The management representative is more likely to rely on the organization's varied expertise in order to accomplish all the energy activities described in this standard. Of course, these assignments should be made with the authorization and concurrence of the appropriate managers.

2.16.2   *How will this benefit us?*

Recruiting the required talent and effort supports the implementation, maintenance, and improvement of the organization's EnMS, which should lead to improved energy performance.

2.16.3   *How do we implement this?*

Forming an energy management team should be accomplished in a manner that is consistent with the norms of the organization. For example, if formal work agreements signed by the employees, their functional managers, and the project/program manager are the norm, then use them here. If verbal commitments are the custom, then that must suffice, as it will be awkward to seek a written commitment. The idea, however, is to obtain specific commitments from certain individuals based on their expertise, knowledge, or availability to accomplish certain tasks within given time frames.

2.16.4 What evidence do we need?

Documented and signed work agreements are evidence that the management representative has formed an energy management team. These are fine if they are the norm in your organization; however, we do not recommend creating them solely for the purpose of demonstrating that the management representative has sought help with energy management activities. The fact that people within the organization are engaging in energy management activities is evidence that the management representative is drawing on organizational support to accomplish energy management activities.

2.17 Plan and direct energy management activities designed to support the organization's energy policy (4.2.2e of the ISO standard)

> The management representative is required to plan and direct energy management activities designed to support the organization's energy policy.

2.17.1 What does this mean?

As he or she is appointed by top management, the management representative is responsible for acting as an energy program manager of sorts, which includes planning and directing all the activities needed to support the organization's energy policy. This includes all the activities discussed in the ISO 50001 standard. As discussed earlier, although the management representative may not do all the actual work by himself or herself, the buck stops with the management representative.

2.17.2 How will this benefit us?

By engaging in "plan and directing" energy management activities, the representative supports the organization in achieving its energy policy.

2.17.3 How do we implement this?

We recommend thinking of the energy management effort, at least initially, as a formal project. Having defined deliverables to be accomplished by a certain date within a prescribed budget helps create the focus and visibility that will be necessary to implement an EnMS that meets the ISO 50001 standard. Using fundamental program and project management processes and tools will help ensure that the necessary activities are performed in a timely manner.

Spending some time up front to think and strategize can pay considerable dividends in the long run in terms of fewer false starts, a better understanding of the organization's existing EnMS, and improved energy performance. Documenting the results of this up-front planning in a project charter approved by top management ensures that everyone is on the same page to begin with. After the improved EnMS is up and running, subsequent activities tend to be less project-like and more integrated into the way the organization does business.

2.17.4 What evidence do we need?

Any documents showing that the management representative is planning and directing the energy management activities being accomplished could be considered evidence.

2.18 Defining and communicating responsibilities and authorities (4.2.2f of the ISO standard)

> The management representative is required to define and communicate responsibilities and authorities in order to facilitate effective energy management.

2.18.1 What does this mean?

As the organization's designated "energy program manager," the management representative is the person with the most knowledge about what needs to be accomplished. Although top management may be responsible for making the formal assignments, the management representative will be the person to define and convey to the rest of the organization the tasks and responsibilities needed for effective energy management.

2.18.2 How will this benefit us?

Having a clear understanding of the responsibilities and authorities necessary for effective energy management enables top management to make informed decisions about job assignments. This results in involvement at all levels in the organization in order to achieve energy improvement objectives.

2.18.3 How do we implement this?

Your organization may already have a formally defined process for establishing roles, responsibilities, accountabilities, and authorities.

2.18.4 What evidence do we need?

Documented evidence that staff have been assigned responsibilities for defined energy-related activities could include organization charts, statements of roles, responsibilities, accountabilities and authorities, job descriptions, or performance management forms.

2.19 Determining criteria and methods for effective operation and control of the energy management system (4.2.2g of the ISO standard)

> The management representative is required to determine the criteria and methods needed to ensure that both the operation and control of the EnMS are effective.

2.19.1 What does this mean?

It is a positive step forward that the organization now has an EnMS, but how do we really know that it is capable of delivering the desired results? Establishing the criteria for effective operation and control of the EnMS includes setting energy targets, defining requirements, identifying the needed standard operating conditions, and identifying critical success factors. These are the criteria the organization will use to determine if the energy management system is effective in managing energy use (see Sections 4.4.5 and 4.4.6 of the ISO standard).

Once the organization determines the criteria they will use to measure the effectiveness of their EnMS, how do they go about meeting these criteria? The methods for meeting these criteria are the regular and systematic actions undertaken by the organization that are intended to ultimately deliver the expected results. Many of the methods will be prescriptive and described in the energy processes and procedures. Others will be less prescriptive and not formally documented.

2.19.2 How will this benefit us?

Determining the EnMS criteria and methods helps the organization understand whether its energy-related activities are achieving the expected results.

2.19.3 How do we implement this?

First, determine the characteristics needed for success. What equipment, facilities, and material are needed? What competencies are required for specific personnel? Then define the process attributes that will improve energy performance. Define the process inputs, which consist of information and material from other processes. Define the major activities that will utilize these inputs. Next, define the process outputs, the resulting information and material that go to other processes. Finally, measure each process's performance and effectiveness using monitoring and metrics.

2.19.4 What evidence do we need?

You will need documented energy management system records. The specific evidence required is discussed in the respective sections of this book.

chapter three

Energy policy

This chapter describes the ISO 50001 standard for implementing an energy policy (ISO 50001; Section 4.3). We begin by describing the requirements to formulate and communicate an energy policy.

3.1 Commitment to improvement

> The ISO 50001 standard requires that the energy policy state the organization's commitment to achieving improvement in energy performance.

3.1.1 What does this mean?

The energy policy commits the organization to achieving broadly stated operational and performance goals and objectives for its energy management system (EnMS) and defines the means for doing so.

3.1.2 What are the benefits?

A successful policy enables the organization to manage the evolution of its EnMS so as to create clearly defined benefits for the organization in terms of cost-effectiveness, product quality, process efficiency, and regulatory compliance as they apply to its own goals and targets for its respective energy systems.

3.1.3 How do we implement?

Implementation begins with a statement of commitment in a published energy policy document that is communicated throughout the organization. The commitment statement should describe the role of top management in monitoring energy improvement performance. The commitment must also be evident in the processes and procedures pursuant to Chapter 4 (Energy Planning), Chapter 5 (Implementation and Operation), Chapter 6 (Checking Performance), and Chapter 7 (Management Review).

The implementation of the organizational commitment to its EnMS policy should aim to ensure that the energy policy:

1. Conforms to the organization's overall management policy (pursuant to ISO 9001) and its environmental policy (pursuant to ISO 14001) to the extent applicable (assuming the organization has ISO 9001 and 14001 systems).
2. Is operationally integrated with federal, state, and local requirements, as well as private sector professional standards and guides respecting environmental, safety, and health requirements.
3. Conforms with Section 4.2 of this book.

3.1.4 What evidence is needed?

Organizational evidence of commitment to improving the performance of the EnMS is provided in the energy policy and by the documented processes and procedures developed to meet the requirements of Chapters 4–7.

3.2 Criteria

This chapter of the ISO standard requires that top management shall ensure that the energy policy meets the eight criteria reviewed in the following sections.

3.2.1 (1) Policy appropriate to energy use

Policy should be appropriate to the nature of, scale of, and impact on the organization's energy use.

3.2.1.1 What does this mean?
The policy should not require more management effort than is warranted by the value of the energy saved in relation to the scale of investment required and the risk involved in achieving the savings.

3.2.1.2 What are the benefits?
Meeting the policy goals and objectives will enable efficient and effective investment in the EnMS within payback periods acceptable to top management. Payback includes the market value of reduced energy consumption and any other operational cost reductions associated with labor and materials or other costs attributed to changes in EnMS technology and operations.

3.2.1.3 How do we implement this?

Section 4.3 of this book must be applied to ensure that they are appropriate to the nature and scale of and impact on the organization's energy use.

The first step should segment the EnMS into its logical functional subsystems such as:

a. Product realization (manufacturing, assembly, etc.)
b. Indoor climate control
c. Transport of inputs from suppliers and outputs to customers
d. Transport of personnel on-site and off-site

3.2.1.4 What evidence is needed?

Section 4.3 of this book documents the parameters of the EnMS appropriate to the nature and scale of and impact on the organization's energy use. Section 4.6 of this book describes how objectives, targets, and action plans are formulated to ensure that these parameters are documented in the energy plan.

3.2.2 (2) Commitment to improvement

> The energy policy includes a commitment to continual improvement in energy performance.

3.2.2.1 What does this mean?

This specifies that the energy policy should include a statement of intent to continually improve energy performance backed up by the appropriate EnMS processes and procedures.

3.2.2.2 What are the benefits?

The overall benefit will be a well-managed energy system that continually demonstrates its effectiveness and improves its performance over time.

3.2.2.3 How do we implement this?

Management at all levels implements the commitment to continual improvement in energy performance by periodically monitoring, measuring, and assessing energy system performance pursuant to the requirements described in Chapters 6 and 7.

3.2.2.4 What evidence is needed?

The energy policy must contain a statement of commitment to continual improvement. Additional evidence includes its implementation in the records of energy performance pursuant to Chapters 6 and 7 of this book.

3.2.3 *(3) Availability of information and resources*

The energy policy includes a commitment to ensuring the availability of information and resources necessary resources to achieve the stated objectives and targets.

3.2.3.1 *What does this mean?*
It means that the information that management deems essential and relevant to attain environmental objectives and targets defined by its energy policy is fully identified.

3.2.3.2 *What are the benefits?*
This commitment drives management at all levels to focus on continual improvements to the EnMS.

3.2.3.3 *How do we implement this?*
By implementing an EnMS management information system that supports the processes and procedures pursuant to this and the following chapters.

3.2.3.4 *What evidence is needed?*
The commitment to ensure the availability of information and of necessary resources to achieve objectives and targets is represented by a statement in the energy policy and further evidenced in the database corresponding to processes and procedures that meet the requirements of this and the chapters that follow.

3.2.4 *(4) Complying with requirements*

The energy policy includes a commitment to comply with applicable legal and other requirements to which the organization subscribes and which relate to its energy use.

3.2.4.1 *What does this mean?*
It means that the organization fulfills its commitment to comply with applicable requirements governing its energy usage.

3.2.4.2 *What are the benefits?*
Satisfaction of customers, regulators, and other stakeholders along with avoidance of regulatory violations, fines, lawsuits, and the operational disruptions that these can entail are the benefits.

Chapter three: Energy policy 33

3.2.4.3 *How do we implement this?*

By providing an appropriate commitment statement in the energy policy to ensure that applicable legal and other requirements to which the organization subscribes and other stakeholder requirements as specified in Section 4.2 of this book (Legal and Other Requirements) will be met through processes and procedures described in the chapters that follow.

3.2.4.4 *What evidence is needed?*

An appropriate statement in the energy policy as well as records that demonstrate compliance with legal and other stakeholder requirements pursuant to this and the chapters that follow are needed.

3.2.5 (5) *Framework*

> The energy policy provides the framework for setting and reviewing energy objectives and targets.

3.2.5.1 *What does this mean?*

The term *framework* means a well-managed planning process or processes for setting and reviewing energy objectives and targets.

3.2.5.2 *What are the benefits?*

It enables top management to assert control over the performance of its EnMS in accordance with the energy policy.

3.2.5.3 *How do we implement this?*

By outlining a framework for processes and procedures pursuant to Chapters 2 and 7 that will facilitate setting and reviewing energy objectives and targets.

3.2.5.4 *What evidence is needed?*

The evidence includes the documented processes and procedures that implement the framework pursuant to the requirements in the chapters that follow.

3.2.6 (6) *Program supports energy efficiency*

> The energy policy supports the purchase of energy-efficient products and services.

3.2.6.1 *What does this mean?*

It means the policy defines top management requirements for a process that supports the purchase of energy efficient products and services.

3.2.6.2 What are the benefits?
A well-documented and cost-effective purchasing program for energy operations that assures that the organization only pays for services and items of demonstrable benefit to the EnMS and that meet quality requirements, such as those in contractual technical specifications are the benefits.

3.2.6.3 How do we implement this?
By applying the requirements laid out in Section 5.26 of this book (Procurement of Energy Services, Products, and Equipment).

3.2.6.4 What evidence is needed?
The evidence needed includes:

a. Records pursuant to the energy procurement process per Section 4.7, Procurement of Energy Services, Products, Equipment, and Energy.
b. Records of management review of the energy procurement program (see Chapter 7, Management Review).

3.2.7 (7) Communication of policy

> The energy policy is documented, communicated, and understood within the organization.

3.2.7.1 What does this mean?
It means that management and staff throughout the organization understand their respective roles in meeting management's performance expectations. It also means that it is documented and communicated to all affected parties (see Section 5.24, "Communications," of this book).

3.2.7.2 What are the benefits?
The result should be effective teams that communicate horizontally and vertically to ensure continuous improvement of the EnMS.

3.2.7.3 How do we implement this?
The energy policy should be communicated by:

- Providing a clear and concise statement of energy policy with achievable goals and objectives that can be periodically monitored
- Posting the energy policy on the organization's website and internal media
- Having small group meetings to communicate the management intent and logic behind the goals and objectives to staff and expert

personnel involved in all aspects of planning as described in Chapter 4 (Energy Planning) of this book
- Requesting participant comments and suggestions for refining and implementing the policy

See Section 5.24, "Communications," of this book for additional guidance.

3.2.7.4 What evidence is needed?
A policy statement regarding how the energy policy is to be documented, communicated, and understood within the organization. These are evidenced by:

a. A documented energy policy approved by management
b. Evidence that the policy is communicated (e.g., by website, video, newsletters, workplace postings, training, etc.)
c. Evidence that personnel are aware of the energy policy, understand, and apply it (e.g., documented test results following training, interviews, etc.)

3.2.8   (8) Policy updating

> The energy policy is regularly reviewed and updated as necessary.

3.2.8.1 What does this mean?
It means that top management ensures that the energy policy is regularly reviewed and updated as necessary.

3.2.8.2 What are the benefits?
Management is assured that the energy policy continues to reflect its commitment to energy performance improvement and a functional framework for moving forward.

3.2.8.3 How do we implement this?
This is implemented by applying processes pursuant the requirements of Chapters 2 and 7. Chapter 2 defines management's roles and responsibilities while Chapter 7 defines the review process that management applies to all aspects of its EnMS.

3.2.8.4 What evidence is needed?
Records indicating that management has reviewed the energy policy and updated it as desired as well as records of successive revisions are needed.

chapter four

Energy planning

Chapter 4 describes the ISO 50001 requirements for energy planning (ISO 50001, Section 4.4). We begin by describing the requirement to conduct and document an energy-planning process. Figure 4.1 conceptually illustrates the energy-planning process that is described in detail in this chapter. This figure is not intended to represent an actual or specific organizational planning process. Nor is this figure exhaustive as there may be other details specific to the organization or specific circumstances.

4.1 General (4.4.1 of the ISO standard)

4.1.1 Perform and document an energy plan

The organization shall conduct and document energy planning that includes the following: legal and other requirements to which the organization subscribes, energy review, energy baseline, energy performance indicators, objectives, targets, and action plans. Energy planning shall lead to activities to improve energy performance.

4.1.1.1 What does this mean?
It means, at a minimum, that the energy plan includes the specific processes essential to improving energy performance: an energy review process, an energy baseline, defined energy performance indicators, objectives, targets, and actions.

4.1.1.2 What are the benefits?
Benefits include an energy management plan that flows from the policy and an EnMS that performs as planned in satisfying the requirements and expectations of customers, suppliers, regulators, and other external stakeholders.

4.1.1.3 How do we implement this?
The energy management plan must be developed by a qualified manager and supported by a dedicated team of professional experts in the various technologies and methodologies to be considered in the planning process.

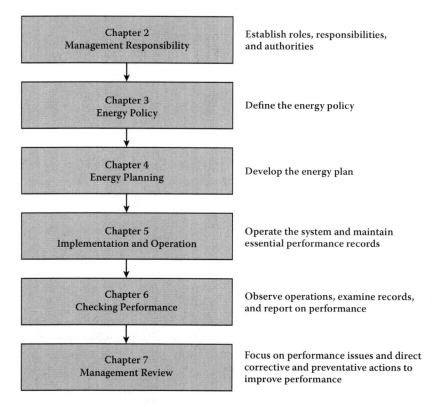

Figure 4.1 Conceptual overview of the ISO 50001 planning process.

Management must evaluate progress at specified milestones in the plan's development and approve the final plan for implementation. Development of the plan should proceed in several stages defined by concrete milestones that are subject to management approval (See Chapter 7 of this book).

The scope includes planning the tasks, subtasks, and resource allocations necessary to address the requirements of the standard.

These activities must be evaluated for their respective inputs and outputs as part of planning to determine which activities need to start only when the previous one is completed and the degree to which certain subactivities within each can proceed in parallel.

Implementation and operation of the EnMS (See Chapter 6) cannot begin until the processes in Section 4.6.6 of this chapter are completed and approved by management.

While the standard does not specifically state requirements for distributing the energy plan, it must obviously be communicated to all persons who will play a role in its implementation.

Chapter four: Energy planning

4.1.1.4 What evidence is needed?

The completed energy plan must provide documented and other evidence of having met the requirements addressed in the balance of this chapter. It should also include the following elements pursuant to developing a Standard Planning Documentation Template:

- Task description
- Milestone schedule chart
- Task-dependency diagram with defined inputs and outputs
- Table of assigned roles and responsibilities
- Progress reporting
- Schedule for management review of planning progress
- Template for documenting direction provided by management review

4.1.2 Activity review

Section 4.4.1 of the standard also states:

> Energy planning involves a review of the organization's activities which can affect energy use and consumption, or relate to them in a wider sense. Having brought this data and information together, a range of tools and techniques are available to develop the energy planning outputs.

4.1.2.1 What does this mean?

It means going beyond other stated requirements by evaluating the energy system in the larger context of the organization's energy demand factors and utilizing tools and techniques that may be available for investigating various aspects of the energy-planning process and outputs that will be generated by this process.

4.1.2.2 What are the benefits?

The benefit is a plan that achieves the organization's needs and opportunities for reducing energy consumption.

4.1.2.3 How do we implement this?

By requiring a larger context for planning and using the best tools and techniques available.

4.1.2.4 What evidence is needed?

The evidence must be provided in the context of the appropriate plan elements, as discussed in Sections 4.2–4.6 that follow.

4.2 Legal and other requirements (4.4.2 of the ISO standard)

4.2.1 Identify and have access to the applicable legal and other requirements

The organization shall identify and have access to the applicable legal and other requirements to which the organization subscribes related to its energy uses. The organization shall determine how these requirements apply to its energy uses and shall ensure that these legal and other requirements to which the organization subscribes are taken into account in establishing, implementing, and maintaining the EnMS.

4.2.1.1 What does this mean?

Legal requirements include but are not necessarily limited to federal, state, and local laws, regulations, and ordinances pertaining to energy production, consumption, environmental factors, workplace safety and health, emergency management, inspections, audits, and documentation. Other requirements may include those that management establishes such as meeting customer, regulator, and other stakeholder interests in the quality of the EnMS, along with compliance with internal business procedures and operations.

4.2.1.2 What are the benefits?

Meeting these requirements should enable an energy management plan that anticipates and responds to requirements, needs, and expectations of all stakeholders (including management and workers). The benefits of such a plan include avoidance of law enforcement actions, court appearances, fines, and hassles, while earning regulator, customer, and other stakeholder confidence in the management of the EnMS.

4.2.1.3 How do we implement this?

While management itself is the principal expert in performing its business affairs, an energy management plan should employ additional legal and technical expertise as warranted to ensure that the large variety of regulatory, customer, and stakeholder requirements, needs, and expectations are addressed in the energy management plan. An assessment of requirements should be organized around each of the following functions that require energy:

(1) Product realization (including waste management processes pursuant to it)
(2) Indoor climate control

Chapter four: Energy planning 41

(3) Transportation of product (energy, materials, etc.) from suppliers to the company and from the company to its customers.
(4) Transportation of personnel onsite and offsite in the performance of their work

All potentially relevant sources of legal requirements and industry standards that may benefit the energy plan and its implementation should be explored, such as:

- U.S. Occupational Safety and Health Administration (OSHA) regulations (and equivalent regulations in other countries) that may apply, along with state and local requirements as may be applicable.
- U.S. environment requirements (and equivalent requirements in other countries) that may apply along with state and local requirements as applicable.
- Compliance with other sources of regulation of operations that are potentially hazardous to the health and safety of persons in the workplace or to the environment (air, water, soils, land) and to its living species, as may be applicable.

The following standards of professional organizations should also be investigated, as they are often required or cited by regulatory requirements, or are otherwise essential or desirable for the design and operation of components of the EnMS:

- American Society of Mechanical Engineers
  Process heating   EA-1-2009 & EA-1G-2010
  Pumping   EA-2-2009 & EA-2G-2010
  Steam   EA-3-2009 & EA-3G-2010
  Compressed air   EA-4-2009 & EA-4G-2010
- American Society of Heating, Refrigerating and Air-Conditioning Engineers (ASHRAE) for various applicable standards
- American Institute of Chemical Engineers (AIChE), Center for Chemical Safety: *Guidelines for Hazard Evaluation Procedures.*
- Institute of Electrical and Electronic Engineers (IEEE) National Electrical Safety Code (NESC)
- National Fire Protection Association - Electrical Code (NFPA #70)
- ISO 14121-1: ISO Safety Standard to Protect Machinery Operators

It must be emphasized that these are only examples of requirements that may or may not apply to a given EnMS and do not constitute a comprehensive list.

National Environmental Policy Act–U.S. National Environmental Policy Act (NEPA) requires that an environmental impact assessment be performed prior to making a final decision regarding a federal action; many other nations have environmental impact assessment processes that have been modeled after NEPA.

For instance, if the organization is a U.S. government entity, or a private entity or state agency pursuing an action such as obtaining a federal permit or authorization, this action may trigger the requirements of NEPA, such as having to prepare an Environmental Impact Statement (EIS) or an Environmental Assessment (EA) to evaluate impacts of the proposed action and to compare them with impacts of feasible alternatives to that action. The former is a relatively long, complex, and hence expensive process. The latter is a shorter process that usually results in a "finding of no significant impact," or "FONSI," that exempts the proposed action from the requirement to prepare an EIS. Two recommended references on how to manage NEPA requirements include the author's companion books:

> Charles H. Eccleston, *NEPA and Environmental Planning: Tools, Techniques and Approaches for Practitioners.* CRC Press 2008. This book is a comprehensive guide to methodologies used in the assessment of environmental impacts, documenting them for public review, and informing the federal decision process.
>
> Frederic March, *NEPA Effectiveness: Mastering the Process.* Government Institutes 1998. This book is a concise guide to the NEPA requirements designed to help the user fully understand the act and the regulations that guide its implementation.

4.2.1.4 What evidence is needed?

Documentation is needed that identifies the applicable legal and other requirements and demonstrates that they have been applied in the planning process by personnel having the requisite training, experience, and qualifications pursuant to those requirements. Moreover, each requirements source specifies additional documentation that may be required for legal or contractual statements.

4.3 Energy review (4.4.3 of the ISO standard)

4.3.1 Recording and maintaining the review

The organization must record and maintain an energy review with a documented methodology and criteria.

4.3.1.1 What does this mean?

An energy review is a process that assesses the organization's current energy performance and identifies opportunities for improvement as an

Chapter four: Energy planning

input to developing the Energy Baseline (see Section 4.4) and the selection of Energy Performance Indicators (Section 4.5)

The review should not be overly detailed and should largely employ readily available data on current costs of energy, the condition of the major equipment that consumes the energy, and the evident needs of the energy management system—physical, organizational, and expertise.

It should target opportunities to be developed in later planning stages such as energy technologies and source substitutions, including renewables, electronic control applications, material substitutions, and replacement of selective system components and logistical considerations.

4.3.1.2   *What are the benefits?*

The energy review enables proceeding to the next steps of the energy planning process with a good sense of target opportunities to explore further. It also establishes the monitoring capability to support effective continuous EnMS improvement in the future. The energy review is designed to provide inputs to guide the energy baseline (See Chapter 4) and the selection of energy performance indicators (See Chapter 5). The energy review report provides a high level of strategic intent based on available data at this early stage in the process. Its principal outputs include a data gap identification to be addressed in Chapters 4 and 5.

4.3.1.3   *How do we implement this?*

Qualified energy engineering auditors assess all relevant documented data about the energy features of all facilities and equipment in the system that are available. They may also prescribe new measurements to fill gaps in the record and to enable the future routine monitoring of energy consumption and related operational parameters. They also may apply models that simulate EnMS performance predictions by varying the parameters of its individual components.

4.3.1.4   *What evidence is needed?*

An energy review report with all analysis and data, along with documentation of management's review and approval of the report is necessary. Its contents and outcomes should include:

- The scope of the planning process, which should include planning how to meet all of the requirements specified in the outputs if applying Chapter 2. Essential gaps of data and related knowledge to be addressed in subsequent stages of the planning process are also identified and documented.
- Statements of strategic purpose and direction to guide the implementation work to be done pursuant this section; these may appear in the Energy Policy (See Chapter 3)

- Specific steps to be taken to produce the evidence of performance against the various requirements in the Plan's Objective's Targets and Action Plans (Section 4.6), Checking Performance (Chapter 6), and Management Review (Chapter 7).
- Clear statements of management expectations for validating and recording performance documented by a stated schedule of progress reports to management.

4.3.2 Analyzing, identifying, prioritizing, and recording

The standard also requires that the following details be implemented:

> To develop the energy review, the organization shall:
>
> (a) Analyze energy consumption based on measurement and other data (4.4.3a of the ISO Standard):
> - Based on energy use analysis, identify the areas of significant energy use and consumption (4.4.3b of the ISO Standard)
> - Identify, prioritize, and record opportunities for improving energy performance (4.4.3c of the ISO Standard)
> - Perform the energy review at defined intervals and in response to major changes in facilities, equipment, systems, or processes (4.4.3 of the ISO Standard)

4.3.2.1 What does this mean?

The required energy review is a three-step process as illustrated in Figure 4.2.

By analyzing current measurements of energy use and past records that reveal patterns and trends, management can assess which uses are most significant to the organization's operations. This enables targeting of opportunities for saving the most energy in Step 3. This three-step process is also used to update energy saving targets when major system changes occur in the future.

The term "significant energy use" means a level of consumption that offers considerable potential for energy performance improvement according to the organization's specific criteria. Table 4.1 illustrates a hypothetical example of scoring the significance of energy use on a simple scale (low, medium, and high) of potentially significance energy consumption.

Chapter four: Energy planning

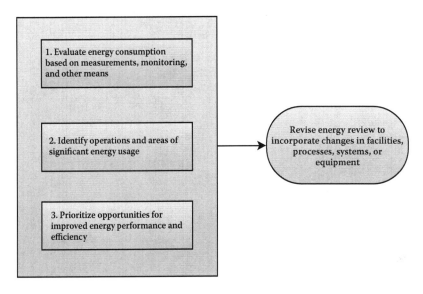

Figure 4.2 Three-step energy review process.

Table 4.1 Hypothetical Example of Significance of Energy Consumption Levels as a Priority Tool

Energy use	Consumption (million kw h/year)	Significance score
Space cooling	1.1	High
Process steam	0.95	High
Space heating	0.6	Medium
Forging the widget	0.3	Medium
Packaging the widget	0.02	Low
Transporting the widget	0.02	Low

While the standard only addresses consumption for various energy uses, other criteria may be important to your organization's priorities. For example, if your operation is subject to a tax or fines on greenhouse gas emissions and/or other atmospheric emissions, the organization may want to apply the criteria as illustrated in the hypothetical example of Table 4.2.

4.3.2.2 *What are the benefits?*

The required three-step process enables management to make initial decisions that become inputs to design of the Energy Baseline (Chapter 4), which in turn influences the choices of Energy Performance Indicators

Table 4.2 Hypothetical Example of Significance of Greenhouse Gas Emissions as a Priority Tool

Energy use	Fuel consumption (million liters)	Emissions (million kg/year)	Cost of emissions	Significance
Diesel used in for transporting widgets	1.0	2.7	$270,000	High
Gasoline used in plant vehicles	5.0	1.2	$120,000	Medium
Heating oil	1.0	0.3	$30,000	Low

(Chapter 5). These initial decisions can be revised when you get to setting the energy system objectives, targets, and action plans (Chapter 6) and yet again after each period of operation in the management review (Chapter 7).

4.3.2.3 How do we implement this?

Analyzing energy use requires several categories of special expertise. Most organizations are likely to need the services of energy audit companies that may specialize in buildings, industrial processes, or transportation. Prior to contracting for energy audit services, it may be helpful to learn about what is involved from sources such as the following:

> Lawrence Berkeley National Laboratories: Industrial Energy Audit Guidebook: "Guidelines for Conducting an Energy Audit in Industrial Facilities," October 2010 (Prepared by the Environmental Energy Technologies Division, Energy Analysis Department, China Energy Group).
> Bonneville Power Authority: A Guidebook for Performing Walkthrough Energy Audits for Industrial Facilities (undated).
> Building Performance Institute, Inc.: BPI-101 Home Energy Auditing, Draft May 2010.

Energy improvement opportunities selected by Management Review should be grouped into manageable projects that include specific milestones, schedules, targets, team composition, and funding. See Chapter 5 for integrating the various project plans.

4.3.2.4 What evidence is needed?

Evidence needed includes the databases, the record of suggestions for improvement, the Assessment/Audit Team reports, and the records of Management Review of the energy review process and its outcomes.

Chapter four: Energy planning 47

4.3.3 Preengineering decisions for baseline

The standard continues:

> (b) Based on energy use analysis, identify the areas of significant energy use and consumption:
> - Identify the facilities, equipment, systems, processes and personnel working for or on behalf of the organization that significantly affect energy use and consumption
> - Identify other relevant variables affecting significant energy use and consumption
> - Determine the current performance

4.3.3.1 What does this mean?

It means obtaining the data essential to making preengineering decisions about the overall EnMS configuration of facilities, systems, processes and the deployment of personnel in EnMS operations as they pertain to energy use.

4.3.3.2 What are the benefits?

This step provides essential information needed to establish the Energy Baseline (See Section 4.4) as well as to specify Energy Performance Indicators (See Section 4.5).

4.3.3.3 How do we implement this?

It means a comprehensive Energy Audit by qualified energy engineering auditors familiar with applicable energy industry standards.

4.3.3.4 What evidence is needed?

An Energy Audit Report that documents the above current energy uses and related operational factors is needed.

4.3.4 Selecting technologies and methods

The standard continues:

> (c) Identify, prioritize, and record opportunities for improving energy performance, including, where applicable, potential energy sources, use of renewables, or alternative energy sources.

4.3.4.1 What does this mean?

It means selecting the technologies and methods that offer the best opportunities for initially improving future EnMS performance.

4.3.4.2 What are the benefits?
Prioritizing opportunities at this stage provides essential inputs pursuant to Section 4.4 (Energy Baseline), 4.5 (Energy Performance Indicators), and Section 4.6 (Objectives, Targets and Action Plans).

4.3.4.3 How do we implement this?
The energy audit team analyzes the data, writes an EnMS Opportunities Report, and helps management decide on its overall EnMS investment strategy.

4.3.4.4 What evidence is needed?
Evidence includes the EnMS Opportunities Report and the record of management decision pursuant to Chapter 7.

4.3.5 Updating reviews
The standard also requires that the energy review should be updated.

4.3.5.1 What does this mean?
Future energy reviews should be planned, scheduled, and budgeted.

4.3.5.2 What are the benefits?
These reviews provide management with critical evaluations about continuing opportunities to improve EnMS performance.

4.3.5.3 How do we implement this?
Top management establishes a schedule for performing updated energy reviews. The experience of operating the EnMS documented pursuant to Chapters 5, 6, and 7 enters into the scheduling decision. In general, updates are likely to be justified in 1–3 years. The decision depends upon changes in the availability of energy and other external factors, as well as on the EnMS operating experience.

4.3.5.4 What evidence is needed?
The record of the decision to update pursuant to Chapter 7 is needed.

4.4 Energy baseline (4.4.4 of the ISO standard)
The standard requires that:

- The energy baseline shall be established using the information in the initial energy review considering a data period suitable to the organization's energy use.

- Changes in energy performance shall be measured against the energy baseline.
- Adjustments to the baseline shall be made when:
- Energy Performance Indicators (EnPIs) no longer reflect organizational energy use;
- There have been major changes to the process, operational patterns, or energy systems; or
- According to a predetermined method, the energy baseline shall be maintained and recorded.

4.4.1 What does this mean?

The energy baseline considers all of the data that represent the current status of the energy system together with a narrative that assesses its general condition and history. This baseline represents the primary performance input to the overall planning process and constitutes an essential input to Section 4.6 (Objectives, Targets, and Action Plans).

Inputs to be considered in defining the elements of the energy baseline should include but are not limited to the following data:

- Descriptions of energy system and its subsystems and of major operations, facilities, and equipment that require local management of their energy inputs.
- Historical records of all forms of energy inputs (purchased or generated electric, gas, and other energy sources) and associated data such as documented and estimated fuel costs, local hours of operation of equipment, and facilities. These must be organized in logical categories and units. For example, by building spaces heated and cooled by specifically metered inputs; by individual industrial processes or units of operation, or in logical clusters; by type of vehicles used to transport people, goods, and materials onsite and offsite.
- Historical records of energy use should also include, when applicable, processes, the management, treatment, and transportation of waste, pollution prevention measures, and procedures or processes vital to the assurance of workplace safety.

4.4.2 What are the benefits?

A well-defined and properly detailed baseline enables decisions in the planning process for all levels of the energy system that are grounded on reliable current energy consumption and associated costs. This enables sound decisions about energy system investments by analyzing the relative benefit–cost implications of alternative planning scenarios in Chapter 6.

4.4.3 How do we implement this?

Create a project team to assemble all relevant records in a database designed to support the planning process. If in-house expertise is not sufficient, then contract an energy systems analysis expert with the required experience.

The Energy Baseline must use the output results of Section 4.3 (Energy Review) as an initial input and ensure that it represents an acceptable approximation of the current pattern of energy system performance over a designated design period such as a calendar year.

Targeted changes to this performance must then be measured against this baseline. If the current energy system has energy performance indicators (See Chapter 5) they need to be evaluated for whether they still reflect the organization's use for various reasons because of changes to the process, operational patterns, or energy systems, or according to a predetermined method.

4.4.4 What evidence is needed?

A quality-assurance and documented database of all essential energy baseline information as described above should be maintained. Consider organizing energy baseline documentation in two parts.

> **Part 1:** A database to be used as an ongoing reference by management and by planning, design, and operational personnel at each stage of the subsequent process steps (Chapters 5 through 7). It should be organized in a systematic and user-friendly manner.
>
> **Part 2:** A summary report to be made available to all relevant organizational personnel as a communication and awareness tool. Its contents should include:
>
> - Introduction: Statement of Purpose and Uses of the Baseline.
> - A table listing each form of energy use for the organization as a whole and breakout tables for specific organizational units and functions as may be required by management.
>   - For each table and its energy-related measures per year (or other time unit) provide:
>   - Baseline energy use
>   - Baseline energy cost
>   - Management target energy use and savings

Chapter four: Energy planning 51

– Management target energy cost and savings

For each listed item, provide a short summary of management's strategy for achieving the respective energy and cost-saving targets.

4.5 Energy performance indicators (4.4.5 of the ISO standard)

The organization shall indentify EnPIS appropriate for monitoring and measuring energy performance. The methodology for determining and updating the EnPIS shall be recorded and regularly reviewed. EnPIS shall be reviewed and compared to the energy baseline on a regular basis.

4.5.1 What does this mean?

Energy performance indicators are tools that enable management to periodically assess energy system performance against expectations established in its plan. Performance indicators typically measure energy use and its efficiency per unit of performance such as:

Energy/unit space to heat and cool buildings
Energy/unit of production of a given machine or process
Energy/unit of materials or people transported
Energy/unit of light delivered to essential functions
Energy/unit of electric power delivered for specific functions

These metrics should aim to provide more detail than the measures provided in the baseline and will allow managers to construct a set of before and after energy consumption rates against the most current baseline—as well as cumulative energy consumption and cost over any calendar period.

This database also allows reliable estimates of the total energy used and its cost for any period of record, starting with a defined baseline. Local managers way wish to implement additional metrics that help them identify ways to trim energy consumption. Ultimately, this means that managers can evaluate the effectiveness of energy saving measures in operations over which they have decision authority.

4.5.2 What are the benefits?

Energy performance indicators enable the design of measures essential to identifying opportunities to reduce energy waste and improve efficiency of energy utilization. These data are used in making planning and design decisions and in monitoring actual performance against targets and goals.

4.5.3 How do we implement this?

Energy indicators are specified by the organization's management supported by internal process experts and by energy system audit experts who together can best determine the indicators most essential to achieving the most cost-effective energy performance improvements. Approaches to improving energy systems performance include:

> Fuel substitution
> Energy technology substitution (wind electric power, solar electric power, solar thermal, geothermal, biomass)
> Material substitutions that reduce energy losses
> Energy transfer mechanisms that capture the energy discharged by one process to become a useful input to another process thereby reducing it use of purchased energy
> Technologies that reduce energy required to transmit and manage electric power, as well as to move gas and liquid fuels in pipelines
> Technologies and methods that reduce consumption: insulation, lubricants, electronic controls, reapplication of waste energy, rescheduling of operations and various innovations. These often depend on the local knowledge, expertise and innovation of engineers and technicians in the various units that consume energy
> Common sense changes to reduce wasted energy through simple energy flow controls activated by timers, more efficient light bulbs, and general energy awareness campaigns that motivate the full participation of all employees.

Examples of the diversity of local energy management metrics include:

- Continuous measurements of temperature variations (from which energy and cost can be computed and summarized in tables) in specific building spaces against a 24-hour, 7-day/week, 4-season temperature target curve determined by the uses of the space for people, equipment, or materials
- Continuous metering of specific operational parameters of manufacturing units that affect energy consumption (such as temperatures, pressures, rotor speeds, vibrations, power draw, voltage, and other operating characteristics that may become targets for improving local energy performance.

What evidence is needed? Documentation of the indicators, the plans to employ them, the records generated by their deployment, and of their use in management review are necessary. (See Chapter 7, Management Review.) The

Chapter four: Energy planning 53

outcomes must be systematically organized to roll up all local performance metrics to the top organizational level. Tables at each level must be clear, consistent, and quality assured.

4.6   Objectives, targets, and action plans (4.4.6 of the ISO standard)

This refers to the activity that actually produces the energy plan that directs management and staff to implement the necessary physical and process changes to the EnMS; it also guides the operations to be monitored as required in Chapter 6.* There are seven parts.

4.6.1   (1) Objectives and targets

The organization shall establish, implement, and maintain documented energy objectives and targets at the relevant functions, levels, processes, or facilities within the organization.

4.6.1.1   What does this mean?

This means that management, having studied the Energy Review Report (Section 4.3) and the Energy Baseline Report (Section 4.4) can confidently establish measurable and achievable energy system objectives and targets for improved performance that can be monitored and documented in progress reports.

4.6.1.2   What are the benefits?

Management now has a set of effective metric tools that can enable them to periodically assess progress achieved and to meaningfully communicate and work with lower levels of staff and management to identify problems, to troubleshoot and truly understand them, and to resolve them.

4.6.1.3   How do we implement this?

Implementation requires a periodic flow of a variety of ongoing communication up and down the management chain. Communication mechanisms include:

A shared database on energy system performance records
Electronic communication of issues, thoughts, and ideas among all participants

---

* The numbers 1–7 that follow have been designated by the authors as a convenience and do not appear in the Standard.

Informal meetings called at any time by managers and key staff at all levels to discuss issues and problems

Formal management review meetings at designated milestone times to review progress and discuss the need for preventive and corrective actions, along with new opportunities to reduce energy consumption and costs and other agenda items pertinent to energy system management such as training, personnel needs, and unforeseen events

4.6.1.4   What evidence is needed?

A written record of each meeting should summarize the issues, approaches and path forward, and be made accessible to all managers and key staff in a dedicated user-friendly database as a vital resource for all managers and key staff. (See Chapter 7)

4.6.2   (2) Measures and time frames

The energy objectives and targets shall be specific and measurable. Time frames shall be established for achieving of these objectives and targets.

4.6.2.1   What does this mean?

It means that management applies the energy performance indicators to periodically take the pulse of the EnMS.

4.6.2.2   What are the benefits?

Management decisions are grounded on a solid understanding of actual energy system performance over time.

4.6.2.3   How do we implement?

This requirement is implemented by the processes required in Chapters 6 and 7.

4.6.2.4   What evidence is needed?

The evidence consists of the records of the metrics and the meetings discussed above.

4.6.3   (3) Consistent targets

Targets shall be consistent with the energy policy. Targets shall also be consistent with the objectives.

4.6.3.1   What does this mean?

It means that in addition to the factors previously cited in establishing energy system performance objectives, management must review the

Chapter four: Energy planning

elements of policy described in Chapter 2 and validate that they are consistent with the energy policy and its objectives.

4.6.3.2 What are the benefits?
The organization's energy policy sets a direction that top management considers vital to the future of its business. By reviewing achievements and setbacks in light of the policy, top management is in a position to evaluate the policy's adequacy in driving improved energy performance.

4.6.3.3 How do we implement?
This requirement is implemented by processes pursuant to Chapters 6 and 7.

4.6.3.4 What evidence is needed?
Records of the metrics and the meetings discussed above are needed.

4.6.4 (4) Application and integration of implementation processes
When establishing and reviewing objectives and targets, an organization shall take into account legal and other requirements, significant energy uses, and opportunities to improve energy performance as identified in the energy review.

4.6.4.1 What does this mean?
It means the application and integration of processes and factors that shaped the implementing processes described in Section 4.2 (Legal and Other Requirements), Section 4.3 (Energy Review), Section 4.4 (Energy Baseline), Section 4.5 (Energy Performance Indicators), as well as the processes and factors discussed in the this section. It also means the application of Chapters 5, 6, and 7.

4.6.4.2 What are the benefits?
By taking a quality-assured approach to designing well-integrated processes to create a robust energy management system, the organization is most likely to achieve or exceed management's expressed intentions.

4.6.4.3 How do we implement?
We implement by designing and implementing energy management processes that creatively address the many requirements of the standard.

4.6.4.4 What evidence is needed?
Various forms of evidence as described in each of the sections of this and other chapters are needed.

4.6.5 (5) Consideration of all factors

It (the organization) shall also consider its financial, operational, and business conditions, technological options, and the views of interested parties.

4.6.5.1 What does this mean?
This means that in implementing the requirements of the standard, top management is to consider all of the factors essential to both the energy management system as well as to the health of the organization's business operations.

4.6.5.2 What are the benefits?
The principal benefit will be a well-managed energy system that fulfills the specifications provided in Chapter 2, and Section 4.6 (Objective, Targets, and Action Plans), in ways that contribute to the success of the organization's business operations.

4.6.5.3 How do we implement?
We implement by designing and implementing energy management processes that creatively address the many requirements of the standard.

4.6.5.4 What evidence is needed?
The various forms of evidence as described in each of the sections of this and other chapters are needed.

4.6.6 (6) Action plans

The organization shall establish, implement, and maintain energy management action plans for achieving its objectives and targets. The energy management action plans shall include:

a. Designation of responsibility
b. Means and timeframe by which individual targets will be achieved
c. Statement of the method by which an improvement in energy performance is verified
d. Statement of the method of verifying the results of the action plan

4.6.6.1 What does this mean?
The above requirements are essential elements of any planning process that is grounded on the realities of the system environment that it seeks to alter.

Chapter four: Energy planning

4.6.6.2   What are the benefits?
A realistic plan is likely to be successful.

4.6.6.3   How do we implement?
We implement by ensuring a clear line of sight to the Energy Policy (Chapter 4) and faithfully implementing the requirements in each of the previous sections of this chapter. It is absolutely essential that the management representative who directs the energy system has the necessary leadership and communication skills and a proven track record of leading teams to achieve successful outcomes in complex and challenging programs and projects.

4.6.6.4   What evidence is needed?
Mainly the evidence already supporting the performance of each previous state of the planning process described in this chapter is needed.

4.6.7   (7) Scheduled documenting and updating

The energy management action plans shall be documented and updated at defined intervals.

4.6.7.1   What does this mean?
These requirements stem from long-term management experience in planning that guides the well-established cycle of "plan, act, monitor, review, and change." Methods for doing so are well established in the literature including ISO 9001 Quality Management Systems Requirements. Action plans must include staffing with technically qualified persons, along with orientation sessions and specialized training that enable optimal performance. Changes to key elements of any action plan such as schedules, budgets, and strategies must be considered and approved by management per the processes in Chapter 7.

4.6.7.2   What are the benefits?
The benefits are action plans that have the best chance of success in achieving their stated objectives and targets pursuant to the organization's Energy Policy.

4.6.7.3   How do we implement this?
Assign technically qualified teams led by well-qualified managers for each action plan, who will provide periodic performance reports to management review.

4.6.7.4   What evidence is needed?
Documented team qualifications, action plans, periodic progress reports, and records of management review at each milestone or more frequently, as circumstance may require are needed.

chapter five

Implementation and operation

This chapter describes the requirements underlying implementation and operation of the ISO 50001 standard (Section 4.5 of the ISO 50001 standard). We begin by describing the requirement to use energy management action plans resulting from the planning process to implement and operate an ISO 50001 Energy Management System (EnMS).

5.1 General (4.5.1 of the ISO standard)

The organization shall use the energy management action plans resulting from the planning process for implementation and operations.

5.1.1 What does this mean?

This means that implementation and operations are grounded on plans that have been developed, authorized, funded, and staffed by top management. The organization is required to conduct and document energy planning that addresses a number of legal and other requirements. The resulting energy plan will include processes for energy review, developing an energy baseline, and defining energy performance indicators, objectives, targets, and action plans that will lead to activities for improving energy performance. See Chapter 4 for more detailed information.

5.1.2 How will this benefit us?

Management is assured its energy performance improvement actions are based on its energy policy and planning.

5.1.3 How do we implement this?

This requirement is enabled by performing the required energy planning process described in Section 4.4 of the standard and Chapter 4 of this book. These energy planning results are implemented and operated according to the requirements of Section 4.5 of the standard and Chapter 5 of this book. Briefly, the organization must:

- Develop an organizational energy policy
- Perform energy management planning which promotes the energy policy

- Use the planning results to develop energy management plans with specific activities designed to improve energy performance
- Implement the improvement activities and operate according to the energy management plans

5.1.4 What evidence do we need?

Planning, implementation, and operating records required in Sections 4.4 and 4.5 of the standard (Chapters 4 and 5 of this book) will be needed.

5.2 Competence, training, and awareness (4.5.2 of the ISO standard)

The ISO 50001 standard requires an organization to:

1. Ensure any person or persons working for, or on its behalf related to significant energy uses are competent on the basis of appropriate education, training, skills, or experience
2. Identify training needs associated with the control of its significant energy uses and the operation of its EnMS
3. Provide training or take other actions to meet these needs
4. Maintain associated records.

5.2.1 What does this mean?

Competence means that personnel possess the required skills, knowledge, qualifications, and capacity to perform their duties that can significantly affect energy use or the implementation of the EnMS. People gain competency through a combination of education and training, coupled with experience and unique skill sets. Organizations should assess their personnel to identify areas where they lack energy-related capabilities, identify training opportunities for acquiring the capabilities, and provide the training. The relevant human resources and training records should be maintained.

5.2.2 How will this benefit us?

A capable workforce is essential in successfully implementing the organization's EnMS and achieving improved energy performance.

5.2.3 How do we implement this?

The organization determines what energy-related job competencies will be needed for employees to have a meaningful understanding of the EnMS,

Chapter five: Implementation and operation 61

the processes and procedures to which they must adhere, and the role of technologies with which they work. These competencies should be integrated within the organization's human resources processes. Employee energy performance should also be part of the employee performance management process.

5.2.4 What evidence do we need?

Provide documented personnel job descriptions that include energy-related competencies, personnel records describing the employees' competencies, and records of employees' training and education. Records assessing the employees' competencies based on those required for their job description should also be documented and maintained.

5.3 Awareness of the energy policy, the energy management system, and procedures (4.5.2a of the ISO standard)

The ISO 50001 standard requires an organization to "ensure that persons working for or on its behalf are and remain aware of the importance of conformity with the energy policy, procedures, and with the EnMS requirements."

5.3.1 What does this mean?

Management is responsible for making employees aware of these energy related requirements and understanding their importance. Employees are expected to carry out their assigned tasks consistent with the defined energy policy and procedures.

5.3.2 How will this benefit us?

Awareness and understanding of policy and procedures will help employees operate effectively, and avoid outages, accidents, and violations of regulatory requirements that can result in shutdown, fines, or findings of negligence. Because the EnMS and its procedures provide the means by which the energy policy is fulfilled, the organization benefits when everyone is on the same page and working towards the same goals.

5.3.3 How do we implement this?

Integrate energy-related process and procedure requirements into the organization's command media so that employees understand that these are not just good ideas or options to consider, but rather organizational requirements. Employees should be able to access the most current versions of these requirements while performing their work. In addition, management should require that energy operating personnel periodically demonstrate their understanding of what the EnMS system requires of them. This should also be done whenever significant changes are made to equipment, processes, or operations that could affect energy performance, the environment, or safety.

5.3.4   What evidence do we need?

Documented command media, which includes the energy policy, the EnMS processes and procedures, and any website and workplace postings, as well as records of training and job performance assessments to demonstrate awareness.

5.4   Roles, responsibilities, and authorities (4.5.2b of the ISO standard)

ISO 50001 requires an organization to ensure that persons working for or on its behalf are and remain aware of their roles, responsibilities, and authority in achieving the requirements of the EnMS.

5.4.1   What does this mean?

All personnel should understand their individual functions, obligations, and power to resolve issues in order to meet the requirements of the organization's EnMS.

5.4.2   How will this benefit us?

When everyone does their part, implementing the EnMS leads to improved energy performance.

5.4.3   How do we implement this?

Meeting this requirement first implies that the organization has defined and documented the appropriate roles, responsibilities, and authorities with respect to their EnMS. Next, these should be communicated to all personnel so they understand their individual roles. This should be repeated whenever significant changes are made to operations that could affect energy performance or safety.

Chapter five: Implementation and operation   63

5.4.4   What evidence do we need?

Evidence includes documented definitions of energy-related personnel roles, responsibilities, and authorities. Management must maintain documented records of personnel competencies (such as training, certifications, and on-the job performance) for these energy-related assignments. These records are also essential for supporting ISO 50001 registration. The auditor will ask several personnel about their individual roles or responsibilities. An affirmative and knowledgeable response is evidence that the organization has taken efforts to make their personnel aware.

5.5   Benefits of improved energy performance (4.5.2c of the ISO standard)

The ISO standard requires organizations "to ensure that persons working for or on its behalf are and remain aware of the benefits of improved energy performance."

5.5.1   What does this mean?

Organizations must communicate with their personnel about the importance and benefits of improved energy performance.

5.5.2   How will this benefit us?

Personnel who understand the context of the EnMS and the benefits of improved energy performance will be more committed to helping the organization reach its goals.

5.5.3   How do we implement this?

Communicate with personnel and provide awareness training about the benefits of improved energy performance. In addition to the communication methods already mentioned, periodically apply the requirements of Chapter 6 to ensure that energy-related personnel understand the importance of their respective roles in capturing the benefits of energy system maintenance and operations.

5.5.4   What evidence do we need?

Provide documented communications (e.g., e-mails, flyers, signs) and training records.

5.6 Individual impacts and contributions to improved energy performance (4.5.2d of the ISO standard)

ISO 50001 requires an organization to ensure that persons working for or on its behalf are and remain aware of the impact, actual or potential, with respect to energy consumption, of their activities and how their activities and behavior contribute to the achievement of energy objectives and targets, and the potential consequences of departure from specified procedures.

5.6.1 What does this mean?

This is where the awareness and understanding cultivated by the organization culminate in commitment and action from its personnel. Personnel should understand how much energy is consumed in the conduct of their individual job duties and how this impacts the organization's ability to achieve its energy objectives and targets. Personnel should understand the energy performance consequences to the entire organization if they do not follow specified energy procedures.

5.6.2 How will this benefit us?

An EnMS adhered to by all employees helps the organization achieve its intended energy objectives and targets and has the potential to improve energy performance.

5.6.3 How do we implement this?

Provide targeted personnel energy training so that all personnel are aware of their energy impacts on the organization. A few ideas for doing so include the following.

- Generate a written training plan that addresses new employees, as well as providing refresher training, and specialized training for those whose jobs directly affect energy use and consumption.
- Develop a general training manual for use by all employees.
- Develop a training video that can be viewed by employees on their computer.
- Design the training by topics and levels such as general energy awareness for all employees, and detailed energy compliance for those whose jobs directly affect energy consumption and use.

5.6.4 What evidence do we need?

Documented communication and energy training records are evidence that the organization took action to ensure employee awareness. For purposes of ISO 50001 registration, the auditor may speak to several employees to test for awareness and gauge the effectiveness of the communication and training activities.

5.7 Documentation requirements (4.5.3.1 of the ISO standard)

ISO 50001 requires organizations to establish, implement, and maintain information, in paper or electronic form, to describe the core elements of the EnMS and their interaction.

5.7.1 What does this mean?

Organizations are required to write down information that describes the core elements of their EnMS. Establishing information means that as elements of the EnMS are defined, they are documented. Many of the initial EnMS planning activities constitute establishing information. Exercising the EnMS will result in plans, schedules, data and reports, all of which must be documented. All of these documents must be maintained according to the requirements of the organization and this standard. In today's environment, most documents are prepared using word-processing, graphics, databases, and other computer programs. The ISO 50001 standard allows for these documents to be maintained in the form of an electronic file or in hardcopy paper form.

The degree of documentation for an organization should be based on consideration of the scale of the organization, the types of activities performed, the complexity of the processes and their interactions, and the competence of personnel. Some organizations use process flow diagrams, energy flow diagrams, and energy assessment protocols to ensure effective energy planning. Many organizations use work instructions and checklists for facility and equipment maintenance to ensure effective operational control.

5.7.2 How will this benefit us?

Documenting where the organization is going in terms of improved energy performance and how it will get there in terms of its EnMS keeps the organization aligned and working towards common energy objectives and targets, which leads to improved energy performance. Documentation provides information and supporting evidence of the effectiveness and efficiency of the EnMS. Documentation, also

- Helps take information out of particular individuals' heads and makes it accessible for others to use
- Helps with training and communication
- Provides evidence for top management, customers, regulators, and other stakeholders
- Keeps everyone from having to figure out the same things for themselves

5.7.3   How do we implement this?

As the standard says, information can be established and maintained in electronic files or paper copies. Furthermore, paper documents may be typewritten or handwritten, as long as they remain legible, readily identifiable, and can be controlled by the organization. Some organizations use internal websites to store electronic files of documents, and others use file cabinets to store paper copies. Either or both are appropriate, depending on the needs of the organization.

5.7.4   What evidence do we need?

Provide electronic files or paper copies of documents that describe the core elements of the EnMS and their interaction.

5.8   Documenting scope and boundaries (4.5.3.1a of the ISO standard)

The ISO standard requires the scope and boundaries of the EnMS be documented.

5.8.1   What does this mean?

Section 4.2.1d of the ISO 50001 standard requires the top management to identify the scope and boundaries of the organization's EnMS. This must also be documented.

5.8.2   How will this benefit us?

Employees and auditors will best be able to determine the scope and boundaries of the organization's EnMS if they can read about it or look at

Chapter five: Implementation and operation 67

a picture. Scope and boundary descriptions that are documented remain static as appropriate and do not change from person to person, based on individual memory and the passage of time.

5.8.3 How do we implement this?

These should be documented and controlled per the organization's document control requirements and ISO 50001 standards.

5.8.4 What evidence do we need?

An official, controlled document describing the current scope and boundaries of the organization's EnMS is needed.

5.9 Documenting the energy policy (4.5.3.1b of the ISO standard)

The standard requires that the energy policy be documented.

5.9.1 What does this mean?

You will recall from Section 4.2.1 of the standard that top management is responsible for establishing, implementing, and maintaining the organization's energy policy. Because the energy policy is the driver for implementing and improving an organization's EnMS and its energy performance, it is paramount that it be documented.

5.9.2 How will this benefit us?

Documenting the energy policy facilitates its dissemination so it can be used as a driver to manage organizational behavior.

5.9.3 How do we implement this?

Write the energy policy down and maintain the documents according to the organization's document control requirements.

5.9.4 What evidence do we need?

Provide electronic files or paper copies of the organization's current energy policy statement.

5.10 Documenting energy objectives, targets, and action plans (4.5.3.1c,d of the ISO standard)

ISO 50001 requires organizations to document the energy objectives, targets, and action plans for achieving them.

5.10.1 What does this mean?

As part of the energy planning process, organizations are required to develop energy objectives and targets, as well as action plans for achieving them. These must all be documented and controlled.

5.10.2 How will this benefit us?

The organization will have a clear understanding of what actions it is trying to implement and be able to refer to the energy objectives and targets later on to measure whether they are being met.

5.10.3 How do we implement this?

Write down the energy objectives, targets, and the corresponding action plans for achieving them. Maintain and control these documents according to your organization's control of documents procedure.

5.10.4 What evidence do we need?

Provide documents describing the organization's energy objectives, targets, and action plans for meeting them.

5.11 Maintaining documents considered by the organization to be necessary for ensuring planning, operation, and control (4.5.3.1e of the ISO standard)

ISO 50001 requires organizations to maintain the documents they consider necessary for ensuring planning, operation, and control.

5.11.1 What does this mean?

In addition to the documents required by the standard, an organization should determine what other documents it requires to ensure effective planning, operation, and control.

Chapter five: Implementation and operation

5.11.2 How will this benefit us?

It will provide more effective planning, operation, and control of the EnMS.

5.11.3 How do we implement this?

Write down the information, create the documents, and maintain them according to the control of documents requirements in Section 4.5.3.2 of the standard.

5.11.4 What evidence do we need?

Accessible EnMS documents are needed.

5.11.4.1 Types of energy documents
- EnMS description (scope and boundaries)
- Energy policy
- Energy planning documents
- Energy review
- Energy objectives, targets, and action plans
- Plans for achieving the energy objectives and targets
- Internal and external documents considered by the organization to be necessary for ensuring planning, operation, and control.
- The organization can develop any documents it determines necessary to effectively demonstrate energy performance and the EnMS.

5.12 Control of documents (4.5.3.2 of the ISO standard)

The standard requires organizations to control the documents required by this standard and the organization's EnMS, including technical documentation.

5.12.1 What does this mean?

Although the standard requires specific documents, it is up to the organization to determine the full set of documents required for effectively operating and controlling its energy processes. Controlling the documents means that they are systematically reviewed, approved, and updated so that personnel can find them when and where they need them, and have some way of knowing they are using the correct version of the document.

The requirements for controlling documents are described in detail below. Documentation can be conceptually represented as follows:

Level 1: The EnMS manual. This manual summarizes the EnMS and steers the reader to more detailed procedures and documents.
Level 2: Organization-wide procedures
Level 3: Facility-or division-specific work procedures
Level 4: Records

5.12.2 How will this benefit us?

Documents help the organization communicate its intent and ensure that energy-related activities are performed consistently and according to requirements. Using documents helps provide repeatability and traceability of energy improvement activities, are instrumental in evaluating the effectiveness and continuing suitability of the EnMS, and provide objective evidence. Controlling documents ensures they are correct and that the most recent and valid version of each document is readily identifiable and available.

5.12.3 How do we implement this?

Start by identifying the types of documents the organization uses for the EnMS. Some examples include:

- Documents that provide information, both internally and externally, about the organization's EnMS
- Energy plans that describe how the EnMS is specifically applied to the organization
- Specification documents that state requirements
- Energy guideline documents that provide recommendations or suggestions for improving energy performance
- Requirement documents that provide information about how to perform activities and processes consistently. Examples include processes, procedures, work instructions, checklists, and drawings.
- Documented records that provide objective evidence of activities performed or energy results achieved.

The organization should establish, implement, and maintain procedures for controlling its documents. Although the ISO 50001 standard does not formally require a documented procedure for doing so, practically speaking, it is difficult to do otherwise. The ISO 9001 and ISO 14001 standards do require documented procedure for how organizations control their documents, and for good reason. It will be much easier for employees to meet requirements if they can refer to a document rather than memorize them. Here are a few guidelines for establishing an organization's documents.

- Begin by under-documenting the processes and procedures. Although some specific documents are required by the standard, improving energy performance is not intended to be a documentation or paperwork exercise. As an organization gains experience, it will have a better idea of what additional documents it requires.
- Involve the employees who will use these documents in the document development process.
- Place the documentation on the intranet or Internet to facilitate quick access to the most recent versions by the employees.
- It is up to the organization to determine how much and what documentation is required and the media to be used, depending on the type and size of the organization, the complexity and interactions of the processes, the complexity of products, customer requirements, the applicable regulatory requirements, the ability of personnel, and the extent to which the organization needs to provide evidence that it is meeting EnMS requirements. Table 5.1 provides a suggested five-step process for developing the EnMS Document Control System.

Table 5.1 A Suggested Five-step Process for Developing an EnMS Document Control System

1. *Determine the document control requirements*: Identify the types of information that will need to be controlled, how these documents will be distributed, and who will need access to them. Existing document control processes should be reviewed. Users of the system should be interviewed to assess their problems and experiences with controlling documents. Describe how documents and revisions will be maintained.
2. *Develop a template*: Design a template to house information, procedures, and data.
3. *Develop document control procedures*: Develop procedures for creating EnMS documents. Specify who will review and approve the documents. Specify how the documents will be distributed and used in training. Describe how existing data and information will be converted into formats appropriate for use in this control system.
4. *Implement a document control system*: Implement an established system for controlling the documents.
5. *Periodically review the process*: Review the process on a periodic basis to ensure that documents are being appropriately controlled and to look for opportunities to improve the system.

5.12.4 What evidence do we need?

A documented procedure specifies how the organization controls its energy-related documents and how it compiles energy-related documents that meet the requirements of this standard.

5.13 Approving documents prior to issue (4.5.3.2a of the ISO standard)

The ISO standard requires organizations to establish, implement, and maintain procedures for approving documents as adequate prior to being issued.

5.13.1 What does this mean?

The organization should decide how they will ensure that the appropriate people review the organization's energy-related documents and approve them for initial use.

5.13.2 How will this benefit us?

Documents are issued that contain the accurate and reliable information necessary for employees to effectively perform their work assignments.

5.13.3 How do we implement this?

Document the approval requirements in the organization's control of documents procedure. Many documents contain an approval signature page. Some organizations retain approval forms, emails, or other evidence of approval in a central location for easy retrieval.

5.13.4 What evidence do we need?

- Control of documents procedure with approval requirements
- Documented approval of energy-related documents

5.14 Periodic review and update (4.5.3.2b of the ISO standard)

The ISO standard requires organizations to establish, implement, and maintain procedures for periodically reviewing and updating its energy related documents as necessary.

Chapter five: Implementation and operation 73

5.14.1 What does this mean?

Organizations should decide how, how often, and under what circumstances to review and update their energy-related documents.

5.14.2 How will this benefit us?

Ensuring that energy-related documents have the most accurate factual information leads to informed decisions and strategies for improving energy performance.

5.14.3 How do we implement this?

Document the requirements in the organization's control of documents procedure. Many organizations use review and approval forms for updating their documents.

5.14.4 What evidence do we need?

Document review and approval requirements in the control of documents procedure. Document evidence of review and approval of the organization's energy-related documents.

5.15 Identifying changes and current revision status (4.5.3.2c of the ISO standard)

ISO 50001 requires organizations to establish, implement, and maintain procedures for identifying document changes and their current revision status.

5.15.1 What does this mean?

Employees must be able to know they are looking at the most current and official version of a particular document. They should also be able to easily determine what changes have occurred from the most recent version to the most current version.

5.15.2 How will this benefit us?

Employees will use the most current documents to help them accomplish their work in the correct manner.

5.15.3 How do we implement this?

Develop a numbering or letter code, often called a *revision* level, to be placed on each document. The revision level could be on each page or on a document coversheet. As a document is updated and revised, its revision level is changed to reflect the new version. We recommend adding a header or footer to your electronic documents warning that if the employee is reading a hardcopy version of the document, it may not be the most current version. This will prompt the employee to access the document repository and check for updates.

5.15.4   What evidence do we need?

Documents that include a revision level code are needed. For the purposes of ISO 50001 registration, upon reviewing a document, the auditor will ask the employee, "How do you know that this is the most current version of the document?" The employee should know it is the most current version either because of the revision level or because it was retrieved from a central repository.

5.16   Ensuring that relevant versions of applicable documents are available at points of use (4.5.3.2d of the ISO standard)

The standard requires organizations to ensure that relevant versions of applicable documents are available at points of use.

5.16.1   What does this mean?

This means that employees must be able to find the right documents when and where they need them.

5.16.2   How will this benefit us?

Personnel can do a better job and adhere to requirements if they have the information they need.

5.16.3   How do we implement this?

Maintaining certain documents in a central repository such as a website, a dedicated electronic folder, or a file cabinet ensures that employees can always locate the most current version of each document. For many product realization documents, such as manufacturing procedures, work instructions, and checklists, it is more useful to have the relevant documents located directly at the work site where they are needed.

5.16.4 What evidence do we need?

Evidence is demonstrated when employees can easily access the relevant documents needed to accomplish their work and they know they are working with the most current versions.

5.17 Ensuring documents are legible and readily identifiable (4.5.3.2e of the ISO standard)

ISO 50001 requires that organizations ensure that documents remain legible and readily identifiable.

5.17.1 What does this mean?

Legible documents means that the information in the organization's documents can be looked at and easily read. This may not be a major issue with today's prevalent use of electronic documents and printers. Readily identifiable means that the readers should easily be able to determine what document they are reading and distinguish it from other similar documents.

5.17.2 How will this benefit us?

Employees accomplish their work more effectively when they can rely on documents that convey the information they were originally intended to. Employees are more likely to be able to locate documents and use the correct ones if the documents are readily identifiable.

5.17.3 How do we implement this?

Prior to approving and publishing any document, review it to ensure that the text and illustrations are readable. In today's business environment, this is really not much of an issue with the use of computers, printers, and photocopiers. Many documents which can be accessed electronically via computer have the added value that they can be magnified many times so that small text and detailed graphics can be read. Any documents containing handwritten dates and signatures should have the information printed directly under the handwriting. Documents not used in an office setting may need additional protection against on-the-job dirt and grease or crumpling and tearing from frequent handling. Some protection methods include lamination, plastic protector sheets, folders, and binders, just to name a few.

Documents can be made easily identifiable by ensuring that they contain information that helps employees know what document they are reading. Examples of document identifiers include:

- Name of the organization
- Document title
- Document classification: policy, process, procedure, work instruction, research report, etc.
- Document date
- Version number
- Unique document identification number
- Author

Many electronic document management systems have document identifiers built into their processes. If your organization is ISO 9001 or ISO 14001 certified, your organization already has procedures for ensuring documents are readily identifiable, so use your existing procedures.

5.17.4 What evidence do we need?

Legible documents are required that contain enough identifier information to ensure employees and auditors know exactly what they are looking at.

5.18 Controlling external documents (4.5.3.2f of the ISO standard)

The ISO 50001 standard requires that organizations ensure that documents of external origin determined by the organization to be necessary for the planning and operation of the EnMS are identified and their distribution controlled.

5.18.1 What does this mean?

External documents are those that are created from outside the organization and impact the way the organization manages its EnMS processes, procedures and energy performance. External documents typically consist of public documents and proprietary documents. Examples of public documents which are widely available include the ISO 50001 standard, regulatory regulations, and published books. Examples of proprietary documents that are customer-or supplier-owned include drawings, specifications, purchase orders, and requirements documents. Controlling these documents means that that their location is known and that only authorized employees are using the relevant version.

5.18.2 How will this benefit us?

As appropriate, energy management activities will be based on accessible and relevant information from external documents.

5.18.3 How do we implement this?

These requirements are implemented much in the same way as internal documents, with a few exceptions. Because someone else created these documents, your organization has no control over the review and approval process used to create the document, the document identifiers used, or the revision process. What your organization can do, however, is clearly specify which version to use or point the user to the most recent version. Many organizations manage electronic web links that point the user to an electronic copy of the document that is managed by the external entity. In this way, employees can be assured they are using the relevant version as managed by the issuing entity. This is also a good way to control the distribution of proprietary documents within the organization. Some organizations put their own cover sheet on external documents, which contains more complete and meaningful document identifier data. Lastly, the organization can control external documents by removing invalid and obsolete documents as appropriate to lessen any potential confusion.

5.18.4 What evidence do we need?

External documents that contain enough identifier information to ensure employees and auditors know exactly what they are looking at are needed. Auditors will want to see how the organization knows who currently possesses a document, the approval process for changing ownership, if needed, and a demonstration of the ability to locate documents. For purposes of ISO 50001 certification, the organization's designated document control procedure administrator should be able to explain and demonstrate these elements to an auditor.

5.19 Preventing the unintended use of obsolete documents (4.5.3.2g of the ISO standard)

ISO 50001 requires organizations to prevent the unintended use of obsolete documents, and suitably identify those to be retained for any purpose.

5.19.1 What does this mean?

This means that the organization should ensure that only the relevant versions of documents are used. Furthermore, any documents that are retained but no longer in general use should be recognized as such.

5.19.2 How will this benefit us?

Arranging it so that employees can only use the correct version of a particular document increases the chances that their work will fulfill the organization's requirements.

5.19.3 How do we implement this?

The easiest way to accomplish this, if possible, is to dispose of documents as soon as they become obsolete. However this may not be feasible if documents are widely dispersed throughout the organization, an unknown quantity of copies exist, or if previous versions must be retained as an audit trail. If documents are accessed from a central location, perhaps a library or a website, administrators can control which versions are permitted to be accessed.

For hardcopy documents created by the organization, a header, footer, cover page, or watermark can be inserted, reminding readers that they may not be reading the most relevant version and directing them to the primary source. This allows employees to easily check and verify that they are viewing the most relevant version prior to accomplishing their work.

5.19.4 What evidence do we need?

With a document control system that ensures only the most relevant document versions are utilized, users will be able to trust their evidence as current.

5.20 Operational control (4.5.4 of the ISO standard)

ISO 50001 requires an organization to identify and plan those operations that are associated with its significant energy uses and that are consistent with its energy policy, objectives, targets, and action plans to ensure that they are resourced and carried out under specified conditions.

5.20.1 What does this mean?

The operations are the processes and procedures that combine energy consumption with facilities, equipment, systems, processes, and personnel to develop, produce, and deliver products and services to customers. The organization is responsible for determining which operations are associated with its significant energy uses. Operations and activities tend to be energy significant if they:

- Significantly affect energy consumption, cost, or energy-related impacts
- Rely on energy sources with high price volatility
- Experience significant changes in energy use compared to the previous period
- Are instrumental to the organization delivering its products and services
- Offer considerable potential for energy performance improvement.

All the action plan activities discussed earlier in Chapter 4 have to be integrated into the operational areas they are designed to impact. The operational areas will require the necessary resources to be implemented according to the action plans. The remainder of this requirement is a reminder that the energy policy drives the development of objectives and targets that are made actionable through action plans. It is the activities associated with these action plans that should be used when identifying and planning the actual operations.

5.20.2   How will this benefit us?

The planning associated with energy savings and cost control are only as good as the implementation of the action plans. Energy significant operations are implemented and modified according to action plans in order to improve energy performance.

5.20.3   How do we implement this?

As you will recall, the energy planning described in Chapter 4 must be conducted in the context of the organization's operations associated with its significant energy uses. Most of these operations probably currently exist in the organization, some will have to be modified, and others may have to be developed. Examples of some common significant energy use operations include:

- Production and manufacturing processes
- Maintenance operations
- Equipment operation
- Procurement processes
- HVAC systems—controls, ventilation, heating, cooling

- Lighting systems
- Refrigeration systems
- Power systems
- Ancillary systems

These operations can be planned using standard process improvement and implementation tools such as lean/six sigma and project plans. Lean/Six Sigma consists of a set of methodologies for operations process improvement that includes reducing energy costs, improving energy efficiency, and improving overall energy performance. Using these methods helps to identify opportunities for improving energy performance, collect energy data, and implement changes. Since energy data is collected at the beginning of the process, changes in energy performance can be measured and monitored as described in Chapter 6. Planning energy significant operations can be accomplished using the following eight steps:

1. Identify and prioritize opportunities
2. Define a project
3. Document and measure current energy performance
4. Analyze and identify opportunities for improvement
5. Optimize the operation and reduce energy waste
6. Implement and validate operation
7. Measure operational effectiveness and improvements
8. Communicate results to top management

In situations where steps 5 and 6 above require significant resources and effort, it may be useful to use formal project management processes to accomplish operational improvements. The following is a simplified list of project management processes that may be useful. Organizations should consider using the ones that are appropriate to the size and complexity of the operation being planned:

1. Initiate the project
   - Develop a project charter
   - Develop a preliminary project scope statement.
2. Plan the project
   - Determine the project scope
   - Create a work breakdown structure with costs and activities
   - Develop a communications plan
   - Conduct quality planning
   - Conduct risk management planning
   - Form the project team
   - Develop project management plans.

Chapter five: Implementation and operation   81

   3. Execute the project
   4. Control the project
   5. Close the project

5.20.4   What evidence do we need?

Documented records of the activities described above are needed.

5.21   Effectively operating and maintaining significant energy use (4.5.4a of the ISO standard)

The ISO 50001 standard requires an organization to establish and set criteria for the effective operation and maintenance of significant energy use or where the absence could lead to a significant deviation from effective energy performance.

5.21.1   What does this mean?

For all of the facilities, equipment, systems, processes, and personnel working for or on behalf of the organization that significantly affect energy consumption, cost, energy-related impacts, or any other factors considered important to energy performance, the organization should define the requirements for how these will be effectively operated, maintained, and evaluated.

5.21.2   How will this benefit us?

This establishes a clear direction to determine how these operations will be conducted and maintained, and provides a clear standard for evaluating their effectiveness.

5.21.3   How do we implement this?

This is accomplished by documenting the organization's operational processes and procedures, defining their intended outcomes, and developing measures to assess their effectiveness.

5.21.4   What evidence do we need?

Process and procedural requirements and the associated measures of effectiveness described in Chapter 6 are needed.

5.22 Operating and maintaining facilities, processes, systems, and equipment (4.5.4b of the ISO standard)

ISO 50001 requires an organization to operate and maintain facilities, processes, systems, and equipment, in accordance with operational criteria.

5.22.1 What does this mean?

The organization must plan the operations of facilities, processes, systems, and equipment associated with significant energy uses, and actually operate them according to its plans. It is up to the organization to determine how they should be operated.

5.22.2 How will this benefit us?

Because operational plans reflect the organization's energy policy, the operating facilities, processes, systems, and equipment associated with significant energy uses should help the organization realize its energy commitment.

5.22.3 How do we implement this?

Use the planning output from Chapter 4 to develop the needed operational plans. Follow the plans, check performance periodically, and make any necessary changes.

5.22.4 What evidence do we need?

Documented operations plans, operations records, audits, performance measures, and corrective actions are required, as appropriate.

5.23 Communicating operational controls to personnel (4.5.4c of the ISO standard)

ISO 50001 requires an organization to communicate the operational controls to personnel working for and personnel working on behalf of the organization.

5.23.1 What does this mean?

This means that if personnel are to follow operational controls as planned, they must be cognizant of those operations plans. They become cognizant of the operational controls when management communicates the requirements to them.

5.23.2 How will this benefit us?

Personnel will be aware of operational controls, understand them, be committed to following them, and ultimately fulfill the organization's operational control requirements.

5.23.3 How do we implement this?

Communicating operational criteria to the appropriate personnel is really a specific application of energy-related communication in general. Section 5.5 (Communication) provides an overall framework for effective communication and several tools for implementation.

5.23.4 What evidence do we need?

Documented communication records would include procedures, work instructions, memos, performance measures, and assessments of operational effectiveness.

5.24 Communications (4.5.5 of the ISO standard)

This section describes communication requirements.

5.24.1 Communicating internally within the organization

ISO 50001 requires an organization to communicate internally with regard to its energy performance and EnMS as appropriate to the size of the organization.

5.24.1.1 What does this mean?
The organization should convey information to all personnel regarding the organization's EnMS and energy performance. The phrase "appropriate to the size of the organization" means that the organization should communicate this information in the same manner it would communicate other important information.

5.24.1.2 How will this benefit us?
Personnel who are made aware of and understand top management's commitment to improved energy performance are more likely to be committed themselves and take action to improve energy performance.

5.24.1.3 How do we implement this?
Communicating within the organization should be planned and resourced as part of the overall energy planning. We provide a comprehensive framework for communicating to achieve energy performance

improvement with the caveat that it should be tailored according to the size of the organization and its individual needs. Start by using the energy planning output to develop an energy performance improvement communications plan. The communication plan should discuss the following elements:

- Background of the energy-related communication
- Purpose of communicating energy performance and the EnMS
- Assessment of the organization's communications culture
- Stakeholders for improving energy performance
- Goals and objectives of energy performance-improvement activities
- Audiences to receive messages
- Key messages to be communicated
- Communication channels within the organization
- Communication products to be developed
- Communication activities to be implemented
- Evaluation of communication effectiveness

One of the communication products of the energy communication plan should be a communications package that top management can use to deliver their messages. The communication package should include:

- Communications responsibilities
- Overview of the organization's energy performance
- Energy improvement snapshot and briefing notes
- Improvement summary and schedule
- Questions and answers
- Answer the question, "What does this mean to me?"
- Specific actions required from personnel

5.24.1.4 What evidence do we need?

Any of the communication products and activities developed above as well as measures of communication effectiveness are needed.

5.24.2 Commitment, awareness, and understanding of personnel

The organization should ensure commitment, awareness, and understanding of personnel, as appropriate to their level and role (4.5.5 of the ISO standard).

5.24.2.1 What does this mean?

Improving energy performance is accomplished by the people performing their roles in the organization. For personnel to do things differently, effective communication must move them from awareness, to understanding, commitment, and ultimately action.

5.24.2.2 How will this benefit us?

The organization benefits by improved energy performance and an implemented EnMS by committed personnel.

5.24.2.3 How do we implement this?

Effectively bringing personnel from awareness to action starts with developing the business case for improved energy performance. Help employees to understand why improved energy performance is necessary, what the organization must do, and the benefits of success.

Effective communication planning involves developing a communication strategy and specific objectives. Develop a communication guide that includes communication channels, integration with other communications, and a connection to organization goals. Obtain sponsor review and approval prior to implementation.

The communication plan should identify all target audiences directly and indirectly affected by the EnMS. This may involve individuals who are external to the organization as well as those internal personnel. Identify the specific target audience's position on the communication for change continuum (e.g., awareness, understanding, commitment, and action). Based on the target audience's specific roles, do they need to be aware of the organization's energy performance goals? Do they need to understand the EnMS and be committed to fulfilling its requirements? This will help the organization to be very specific in defining the desired actions for specific target audiences. We recommend developing "bulleted" messages for general audiences, as well as specific audiences. Briefing notes and a communications plan map are all useful as well for testing and refining messages through dialogue.

When implementing the energy communication plan, identify the organization's communication channels and select the appropriate mix of channels for each specific situation. Develop a communications timeline, identify baseline metrics, and develop an evaluation plan to assess the communication effectiveness. Other implementation activities include:

- Develop a budget
- Identify/qualify internal and external resources to execute production
- Secure resources
- Develop production schedule
- Write content, design materials
- Review/test materials

- Produce materials
- Distribute and display materials
- Collect data

Creating energy performance awareness of the organization's personnel can be accomplished by:

- One-way communications
- Disseminating information
- Answering who, what, when, where, why, and how
- Briefing information, facts, data

Effective communication channels for generating awareness should utilize written or oral communication without dialogue. Examples include:

- Broadcast e-mail
- Broadcast v-mail
- Brief announcement letters or memos
- Websites
- All-hands meetings
- Heads-up messages (e.g., invitations)
- Company newspapers
- Company daily newsletters
- Company video networks

Helping the necessary personnel understand the importance of improving energy performance can be accomplished by:

- Two-way communications: discussion, dialogue
- Active listening
- Continuous repetition
- Creating shared meaning
- Understanding audience(s) needs, issues, concerns
- Clarifying the message by putting it into audience(s)' frame of reference
- Building on facts
- Maintaining message consistency and addressing contradictions

Communication channels for effectively generating understanding should utilize written or oral communication without dialogue. Several print media and dialogue examples include:

- Detailed letters and memos
- Web content with hyperlinks

- Large group meetings with facilitated dialogue
- Video conferencing with two-way communication
- Interactive training sessions

Creating the necessary personnel commitment to improving energy performance can be accomplished by:

- Creating a clear connection between personnel's behavior, organizational goals, results and rewards
- Encouraging personnel to question and challenge
- Involving those affected by the change in problem-solving
- Providing diverse channels for open communication

Examples of effective face-to-face dialogue communication channels for gaining commitment include:

- Small group dialogue (e.g., department meetings)
- Team meetings
- Feedback forums
- One-on-one meetings
- Resolution of issues via conversation

Creating the necessary personnel action for improving energy performance can be accomplished by:

- Continuous involvement; dialogue, feedback
- Leaders modeling specific communication behaviors
- Recognizing and rewarding desired behavior
- Participative problem-solving
- Ongoing support and resources
- Measuring results and communicating progress

Examples of effectively reinforcing management communications and creating proactive personnel action include:

- "Walk the Talk" and "Model by Example"
- Continuous updates
- Recognition (e.g., success stories)
- Team problem-solving

The final steps of the communication process are to measure the effectiveness of the communication compared to the baseline, report the results to top management, and adjust the communication plan and materials as needed.

5.24.2.4 What evidence do we need?

Use the tools and processes that make sense for the size of your organization. For an organization of ten people, most of the planning might occur in your head, and you will only produce the communication products. For large organizations, communicating energy-related matters may require significant resources, and you will find it useful to produce many of the planning tools described above.

5.24.3 Making comments and suggesting improvements to the EnMS

This shall include a process by which any person working in or on behalf of the organization can make comments or suggest improvements to the EnMS (4.5.5 of the ISO Standard).

5.24.3.1 What does this mean?

The organization must establish and effectively implement a formal process to collect, analyze, and use comments and suggestions from anyone in order to improve the EnMS.

5.24.3.2 How will this benefit us?

The EnMS improves as the organization benefits by learning from the experience of all personnel.

5.24.3.3 How do we implement this?

This requirement can be implemented by methods that range from simply using a suggestion box, all the way to a formal lessons-learned process. The basic steps are as follows:

1. Notify personnel as to how they may contribute energy-related suggestions and comments. This could include personnel external to the organization as well.
2. Provide a mechanism for collection of energy-related suggestions and comments from work activities, process reviews, event analyses, and other sources.
3. Identify comments and suggestions.
4. Document comments and suggestions.
5. Screen comments and suggestions.
6. Implement changes to the EnMS as appropriate.
7. Communicate changes to affected audiences.
8. Evaluate the effectiveness of the lessons-learned process and improve upon it. Ensure ideas are being collected and considered per any frequency, cycle time, or quality expectations the organization has defined.

Chapter five: Implementation and operation

5.24.3.4 What evidence do we need?

The ISO 50001 auditor will look for objective evidence that suggestions and comments are being tracked and used to drive continual improvement of the EnMS. The auditor will review the EnMS processes and records to see how the organization identifies, documents, screens, and evaluates suggestions and comments. The auditor would likely interview operations managers, support managers, and staff.

5.24.4 Decide whether to communicate externally

The organization shall decide whether to communicate externally about its EnMS and energy performance, and shall record its decision. If the decision is to communicate externally, the organization shall establish and implement a plan for this external communication.

5.24.4.1 What does this mean?

In the context of all the communication guidance provided earlier in this section, top management should decide if their target audience includes anyone external to the organization. This decision should be documented.

5.24.4.2 How will this benefit us?

This decision signals to the organization that its communication processes may need to be modified to include external communication. Information being provided outside the organization may require additional reviews and approvals.

5.24.4.3 How do we implement this?

This is implemented using the methodologies and analyses described earlier in the communication section of this chapter. The relevant audiences would include the external parties, if any, with whom the organization has decided to communicate.

External communications can be managed by identifying a single point of contact, such as a public affairs office, for issuing such information. For organizations that do choose to communicate externally, we offer a few ideas:

- Determine the type and level of energy-related information you will make public. Some of this information can be maintained on the Internet.
- Identify the most qualified individual in the office to respond to inquiries, and route it to an appropriate staff member who can be instrumental in formulating a response.
- Ensure that inquiries receive a timely response.
- Maintain records of inquiries and responses.

5.24.4.4 What evidence do we need?
Records documenting the decision of whether the organization will communicate externally about its EnMS and energy performance. If the organization's decision is to communicate externally, an ISO 50001 auditor would look for the types of communication planning and products described earlier in the communication section of this chapter.

5.25 Designing facilities, equipment, systems, and processes (4.5.6 of the ISO standard)

The following section describes equipment, systems, and processes requirements.

5.25.1 Considering energy performance improvement opportunities

ISO 50001 requires an organization to consider energy performance improvement opportunities in the design of new, modified, and renovated facilities, equipment, systems, and processes that can have a significant impact on energy performance. The results of this energy performance evaluation must be incorporated into the specification, design, and procurement activities of the relevant project.

5.25.1.1 What does this mean?
This requirement applies when designing facilities, equipment, systems, and processes that could have a significant impact on energy. Not only should the organization consider the design's basic performance and function, but also any energy performance issues of interest to the organization.

5.25.1.2 How will this benefit us?
Structuring the design process to consider and respond to the energy needs of the organization provides an opportunity to make a positive change that will significantly improve energy performance.

5.25.1.3 How do we implement this?
The first step is to consider the design of the proposed facilities, equipment, systems, or processes. This starts by identifying the relevant design inputs to gather requirements. Examples of design inputs include:

- Needs and expectations of people in the organization, including the operational users and those receiving the output of the process
- Needs and expectations of other interested parties

Chapter five: Implementation and operation

- The organization's policies and objectives
- Relevant statutory and regulatory requirements
- International or national standards
- Industry codes of practice
- Technological developments that could affect future energy performance
- Competence requirements for people performing design and development
- Feedback information from past experience
- Records and data on existing processes and products
- Outputs from other processes
- Operation, installation, and application
- Storage, handling, and delivery
- Physical parameters and the environment
- Requirements for disposal

In order to consider future energy performance in the design process, the standard requires organizations to review the following issues and incorporate any resulting requirements into the design:

- Does this design preclude the use of alternative energy sources?
- What is the right energy source for this design?
- What are the technological options that would meet requirements and offer energy savings?
- Who will maintain this design later?
- How will the existing processes be modified?
- How will the energy baseline be affected?
- Will this lead to sustainable energy improvement opportunities?
- When will these changes affect the EnMS?

As the design progresses through its various stages, conduct one or more design reviews to ensure that the design process is proceeding as expected. The review should consider the following:

- Adequacy of input to perform the design and development tasks
- Progress of the planned design and development process
- Meeting verification and validation goals
- Evaluation of potential hazards or fault modes in product use
- Life-cycle data on energy performance of the product
- Control of changes and their effect during the design and development process
- Identification and correction of problems
- Opportunities for design and development process improvement
- Potential impact of the product on future energy performance

The next step is to verify that the chosen design is capable of meeting the specified requirements from the design review. This is accomplished by checking drawings and specifications, running simulations, and checking that all design requirements have been addressed. It is also a good time to identify any problems and propose necessary actions. Participants should include representatives of the relevant operational areas as well as members of the energy management team. This helps with managing the interfaces between different groups involved in design and ensuring effective communication and clear assignment of responsibilities. To verify the design, one will need to completely understand all of the functional and performance requirements as well as any applicable statutory and regulatory requirements. Be sure that the requirements are complete, unambiguous, and not in conflict with each other.

The final step in the design process is to validate that the design of the resulting product or service is capable of meeting the operational requirements for the specified application, as well as applicable energy requirements. Upon completion, ensure that the energy-performance-related requirements and specifications are retained with the full set of requirements and included in any subsequent procurement activities.

5.25.1.4 What evidence do we need?

Evidence consists of documents and records of design reviews, results of the verification and validation, and any necessary actions. Documented procurement records should reflect the energy-performance-related requirements.

5.26 Procurement of energy services, products, and equipment (4.5.7.1 of the ISO standard)

ISO 50001 requires an organization to inform suppliers that procurement is partly evaluated on the basis of energy performance when procuring energy services, products, and equipment that have or may have an impact on significant energy use.

5.26.1 What does this mean?

Aside from the standard terms and conditions of purchasing, such as price, payment, delivery, inspection, and acceptance, organizations should establish energy-related criteria that will be used to evaluate procurements. In addition, organizations should inform prospective suppliers that these energy-related purchasing criteria are now part of the process used to select suppliers.

5.26.2 How will this benefit us?

Procurement can ensure that suppliers are aligned with the organizations' energy policy and objectives. Procuring energy services, products, and equipment represents an opportunity to directly improve energy performance through the use of more energy-efficient products and services. It also represents an opportunity to promote partner relationships with the supply chain and influence their energy behaviors, which may indirectly improve the organization's energy performance.

5.26.3 How do we implement this?

The organization's purchasing department should develop relevant energy criteria based on achieving the organization's energy objectives and targets. These criteria should be included in requests for quotations and proposals.

5.26.4 What evidence do we need?

Evidence would include requests for quotations and proposals containing the energy-related criteria, any other communication with suppliers referencing the energy criteria, and procurement justifications taking the energy criteria into account.

5.27 Assessing products' energy use over time

ISO 50001 requires that organizations define the criteria for assessing energy use over the planned or expected operating lifetime of energy-using products, equipment, and services that are expected to have a significant effect on the organization's energy performance.

5.27.1 What does this mean?

The organization should decide how it will assess the energy use of the products, equipment, and services which it uses as inputs to its processes. This requirement is meant to apply only to those inputs that have a significant effect on the organization's energy performance. It is up to the organization to decide what is significant for them.

5.27.2 How will this benefit us?

Improved energy performance results from energy-efficient inputs.

5.27.3 How do we implement this?

First, identify and list all the sources of energy use in the organization. Then measure their respective energy use for a fixed time period, say, a quarter or a year. Reorganize the list so that it is sorted by energy use, from highest to lowest. It may be that the 20/80 rule applies (i.e., approximately 20% of the sources of energy use may account for nearly 80% of the organization's total energy use). These are the sources of energy use on which to focus energy performance improvement efforts. Next, for the energy uses planned for consideration, decide what principles will be used to evaluate their relative energy use for as long as they are expected to be used. These criteria should form the basis for future analyses of the organization's energy performance.

5.27.4   What evidence do we need?

A documented list of energy-use assessment criteria and any subsequent analyses indicating the criteria that have been used to assess energy performance are needed.

5.28   Consider contingency and emergency situations and potential disasters

ISO 50001 requires an organization to consider contingency and emergency situations and potential disasters relating to equipment with significant energy use and determine how the organization will react to these situations.

5.28.1   What does this mean?

The organization should consider the availability of energy as it relates to equipment with significant energy use that can influence the performance of the organization. The organization should have plans, or contingency plans, to ensure the availability of the needed energy or replacement of equipment in order to prevent or minimize negative effects on the performance of the organization.

5.28.2   How will this benefit us?

The numerous benefits of disaster preparation and contingency planning are intended to save organizations time and money in the event of a disaster. Several benefits include:

- Minimized business interruption, downtime, and periods of low productivity
- Business survival in the event of a serious incident
- Minimized financial impact

Chapter five: Implementation and operation

- Fulfillment of expectations of business partners, shareholders, stakeholders, strategic alliances, insurers, and regulators
- Improved organizational resilience to extreme adverse conditions
- Increased job security, improved productivity, and improved employee morale
- Increased ability to avoid business interruptions and key processes interruption
- Reduced risk of losing productivity

5.28.3 How do we implement this?

Include energy availability in the organization's business continuity management processes and disaster preparation planning processes. Identify the key energy intensive processes and equipment, and analyze the impacts that would result from threats. Use this information to prepare for the worst, and take steps to improve the resilience to failure of the energy significant infrastructure supporting the key processes.

5.28.4 What evidence do we need?

Documented records indicating planning for the considerations described above are needed.

5.29 Procurement of energy supply (4.5.7.2 of the ISO standard)

The ISO standard requires organizations to define energy purchasing specifications as applicable for effective energy performance.

5.29.1 What does this mean?

Procurement is an opportunity to improve energy performance through use of more effective or efficient products and services; it also affords an opportunity to work with the supply chain and influence their energy behaviors. When procuring energy services, products, and equipment that have or may have a significant effect on energy use, the organization must inform suppliers that its procurement will be partly evaluated on the basis of energy performance. The organization must define energy-purchasing specifications as applicable for effective energy performance.

5.29.2 How will this benefit us?

The organization benefits from improved energy performance and a supply chain partner with mutual energy improvement interests.

Table 5.2 Criteria That may be Considered in Developing the Purchasing Specifications for Energy Supply

- Energy quality
- Availability
- Capacity
- Variation over specified time
- Billing parameters, cost
- Environmental impact
- Renewability
- Others as determined appropriate by the organization

5.29.3 How do we implement this?

Criteria must be defined for assessing significant energy use and performance over the planned or expected operating lifetime of energy intensive products, equipment, and services. The criteria depicted in Table 5.2 should be considered in developing the purchasing specifications for energy supply.

5.29.4 What evidence do we need?

Documented procurement records containing energy-related specifications and evidence that the specifications were considered in the procurement process are needed.

chapter six

Checking performance

This chapter describes the requirements for checking the performance of an ISO 50001 Energy Management System (EnMS), which is described in the ISO 50001 standard, Section 4.6.

6.1 Monitoring, measurement, and analysis (4.6.1)

The standard requires the organization to ensure that the key characteristics of its operations determining energy performance are monitored, measured, and analyzed at planned intervals. Key characteristics shall include, at a minimum:

a. The outputs of the energy review
b. Significant energy uses
c. The relationship between significant energy use and consumption, and relevant variables
d. Energy performance indicators (EnPIs)
e. Effectiveness of the action plans in achieving objectives and targets

6.1.1 What does this mean?

The meaning of each of the preceding is as follows:

The outputs of the energy review refer to outputs required for the review (see Chapter 4.3), in which past, present, and future energy uses are characterized.

Significant energy uses apply to the energy review (see Chapter 4.3) requirement such that the preceding data are analyzed to determine the significant energy uses as defined in the Glossary.

The *relationship* means that one must characterize the variables that affect the consumption and cost of each energy use (e.g., electric, gas, diesel, etc.).

Energy performance indicators (see Glossary) are measurements of process parameters such as energy consumption rates and related process characteristics such as temperature, pressure, fluid velocities, mixing rates, known sources of energy losses, and any other factor that has a relationship to the uses and costs of energy consumed.

Effectiveness of the action plans means one must have metrics of effectiveness pursuant to Chapter 4.6 for each objective, target, and action plan.

6.1.2   What are the benefits?

The collective benefit of applying these requirements will be a set of achievable, cost-effective, and accurate measures of energy system performance that management can review and rely upon to help define energy performance improvement actions.

6.1.3   How do we implement them?

The following steps are indicated:

Step 1: Understand the outputs described in Section 4.3 of this book, in which initial energy metrics are gathered and analyzed, and significant energy uses are identified.
Step 2: Understand the outputs described in Chapter 3 (Energy Policy) in which high-level energy targets and goals are set for the organization.
Step 3: Understand the outputs described in Section 4.4 of this book, in which various important management decisions requiring metrics support are identified.
Step 4: Perform the analysis required to determine how to implement desired energy performance indicators that, when measured and monitored, can help management improve performance of the energy management system as a whole.
Step 5: Write a performance metrics plan that has the following elements for each significant performance metric to be considered for implementation.

- Name
- Purpose
- Technical approach
- How to apply the data obtained
- Cost of design, acquisition, installation, and operation
- Estimate of contribution to energy and other operational benefits

Step 6: Include an overall benefit/cost analysis for energy system investment.
Step 7: Obtain management review and approval to implement the energy system investments.

Step 8: Develop and implement the metrics according to the approved plan.

Step 9: Provide appropriate processes and procedures involving management at all levels to implement the preceding requirements pursuant to the following requirement laid out in the following section, identified by "The organization shall investigate and respond to significant deviations in energy performance."

6.1.4 What evidence is needed?

You will need to document the planning and performance of each of the preceding steps. Specifically, this section of the standard requires the following:

> The organization ensures that equipment used in monitoring and measuring of key characteristics, provides data, which is accurate and repeatable.
> Records of calibration shall be maintained.
> The organization investigates and responds to significant deviations in energy performance.
> Results of these activities shall be maintained.
> Monitoring and measurement results of the key characteristics shall be recorded.
> The organization defines and periodically reviews its measurement needs.
> The organization ensures that the equipment used in monitoring and measuring key characteristics provides data, which are accurate and repeatable.
> Records of calibration shall be maintained.
> Results of these activities shall be maintained.

6.2 Evaluation of legal and other compliance (4.6.2)

The standard requires that at planned intervals the organization shall evaluate compliance with legal and other requirements to which it subscribes that are relevant to its energy uses.

6.2.1 What does this mean?

This means that the organization must have a process for evaluating compliance with legal and other requirements identified in Section 4.2 of this book.

6.2.2 What are the benefits?

Management will be able to monitor progress against planned milestones that meet all indicated requirements, especially those related to energy system technical and economic performance. In some cases, this will avoid violation of laws and regulations, as well as lawsuits.

6.2.3 How do we implement this?

We implement this by ensuring that the applicable milestones are included in the management of objectives, targets, and action plans as required in Section 4.4.6 of the ISO standard (Objectives, Targets, and Action Plans).

6.2.4 What evidence is needed?

The standard requires records of performance for management review (see Chapter 7) of planned milestones for all applicable objectives, targets, and action plans. The standard also requires that the records of the results of the evaluations of compliance be maintained.

6.2.5 What does this mean?

This means records should be kept of compliance with requirements specified in Section 4.2 of this book.

6.2.6 What are the benefits?

Meeting the requirements is essential to achieving the planned economic benefits of the management system while avoiding violation of laws and regulatory requirements as well as civil lawsuits.

6.2.7 How do we implement this?

We implement this by maintaining records as required under the standard as a whole and by periodic management review against planned milestones and when significant deviations from planned performance expectations are observed.

6.2.8 What evidence is needed?

The records already described herein are needed.

6.3 Internal audit of the EnMS (4.6.3 of the ISO standard)

The standard requires organizations to perform internal audits of the EnMS at planned intervals.

6.3.1 What does this mean?

The organization should evaluate the EnMS processes, procedures, and implementation to determine if they are appropriate for the organization, if the requirements are being fulfilled, if they are effective, and if they meet the ISO 50001 standard. These first-party audits are conducted by, or on behalf of, the organization itself for internal purposes and can form the basis for an organization's self-declaration of conformity to ISO 50001 or other requirements.

The term "internal" audit as used in the ISO 50001 standard really is meant to imply that the energy audits are being performed because the organization has voluntarily elected to do so and has control over the audit plan, scope, and execution. The organization can use its own employees to conduct audits or hire an outside auditor. The key thing to keep in mind when selecting auditors is to be sure they are objective and impartial to the audit process. This means that for purposes of an independent audit, staff should not audit their own work, processes, or areas for which they are responsible.

6.3.2 How will this benefit us?

Conducting internal energy system audits can benefit the organization by verifying that it is conforming to legal, contractual, and its own requirements. It helps ensure that the organization is actively seeking to implement all cost-effective measures for reducing energy use. They also contribute to the improvement of the EnMS by identifying nonconformities and opportunities for improvement which management should act upon. Internal audits can also help the organization in meeting requirements for ISO 50001 certification.

6.3.3 How do we implement this?

The first step is for top management to authorize the implementation of an energy audit program. The assigned audit program manager will initially be responsible for defining the program objectives and extent, assigning responsibilities, ensuring that necessary resources are available, and defining the audit procedures to be used. As the program is implemented, the program manager responsibilities will include scheduling the audits, evaluating auditors, selecting audit teams, directing audit activities, and maintaining the needed records. These factors can vary and depend on the size, nature, and complexity of your organization, as well as your energy profile. The entire EnMS can be audited at the same time in a single comprehensive audit, or it can be broken into subsystems that are audited separately. This is up to you to decide.

The next step is to conduct the energy audits according to the criteria discussed earlier. To initiate an audit, assign an audit team leader, define the audit objectives, scope, and criteria, select the audit team, and establish initial contact with the people responsible for the processes to be audited. Defining the audit scope up front is critical because it describes the extent and boundaries of the audit, such as physical locations, organizational units, and activities and processes to be audited, as well as the time period covered by the audit. Conducting a successful audit will require the cooperation of the process owners being audited, so it is important to work out any details such as access to documents, personnel, and facilities prior to initiating the audit.

Because much of the audit evidence may consist of information from documents and records, it is considered a best practice to perform a document review prior to starting the on-site audit activities. If the required documentation is inadequate or does not exist, the audit team may be unable to continue with that portion of the audit scope. If the documents themselves indicate that the EnMS does not conform to requirements, then it may be prudent to continue the audit at a later time after the documentation has been improved.

Prior to conducting the actual on-site audit, the team leader should prepare an audit plan, prepare any checklists, forms, or templates to be used, and assign the individual audit team responsibilities. The on-site audit consists of an opening meeting to confirm what is to be audited and how the audit will proceed, collecting and verify information, generating audit findings, preparing audit conclusions, and conducting a closing meeting to present the audit findings and conclusions. Finally, an audit report is prepared, approved, and distributed.

The last follow-up step occurs after the audit has been completed. The audit program may specify that the audit team follow-up with the auditee to verify that any needed corrective, preventive, or improvement actions have been undertaken within the agreed upon timeframe. These follow-up activities could include verifying that the actions have been implemented and validating that they are effective.

When first establishing an EnMS, many companies focus their audits on determining whether or not the system meets the ISO 50001 standard. This is useful early on because ISO 50001 is a good blueprint for the kinds of activities that need to be conducted in order to continually improve energy performance. However, subsequent EnMS audits should really focus primarily on evaluating how well the system is being implemented according to the organization's own requirements and assessing the overall effectiveness of the system based on the organization's energy objectives, targets, and performance indicators. When auditing your EnMS, ask the following questions in relation to each process:

Chapter six: Checking performance 103

- Is the process identified and appropriately defined?
- Are roles, responsibilities, authorities, and accountabilities assigned?
- Are the procedures being implemented and maintained?
- Is the process effective in achieving the organization's desired results?

If an organization is already ISO 9001 or ISO 14001 certified, chances are it already has an audit program that will meet the requirements of ISO 50001. One option is to have the existing audit program add energy performance audits to their scope. This will require that they have the requisite knowledge of energy issues and the organization's EnMS, and may require additional training. Another option is to develop a separate energy audit program that focuses on the EnMS. The same energy knowledge and possible training needs would apply here as well.

EnMS audits conducted by an ISO 50001 registrar as part of the ISO 50001 certification process are not considered internal audits. Organizations should conduct their own internal audits prior to bringing in a registrar. One of the things your registrar will be looking for is whether the organization has conducted any internal audits. After all, nobody knows the EnMS better than the organization, so one should be able to identify the majority of problems internally. The first question an auditor is likely to ask after identifying an obvious system nonconformity is "Why didn't you discover this yourself?"

6.3.4 *What evidence do we need?*

Documented audit schedules indicating when the audits are to be conducted are evidence that the audits are planned. These should be treated as living schedules; for example, if circumstances change so that an audit cannot be conducted as planned, the schedule can be modified to reflect a revised date. It is not acceptable for your ISO 50001 registrar to review an audit schedule only to find that one or more audits were not conducted as planned. Unforeseen events happen, so change the audit schedule if necessary.

Documented audit plans and reports are evidence that internal audits are being conducted and will indicate the actual dates of the audits. Audit plans should clearly state what energy processes and procedures the audit will cover and the rationale for doing so. The audit scope might be based on the results of previous audits, visibility within the organization, or the importance of a particular process for achieving energy performance and improvement.

Documentation of the audit results, including nonconformities, should be maintained according to the organization's procedure for controlling documents and the results reported to top management. The appropriate venue for reporting internal energy audit results to top management is likely to be during management reviews. In this way, management can take action to ensure needed improvements are made. The ISO registrar auditor may also want to see the objective evidence that root-cause analysis is being conducted on any nonconformities (see the next section).

6.4 Nonconformities, correction, corrective, and preventive action (4.6.4 of the ISO standard)

The ISO 50001 standard requires an organization to establish, implement, and maintain procedures for dealing with actual and potential nonconformities, making corrections, taking corrective action, and taking preventive action.

6.4.1 What does this requirement mean?

A fundamental principle of the ISO 50001 standard is that organizations be capable of identifying problems that prevent their energy requirements from being fulfilled, taking action to fix the problem, and taking subsequent action to eliminate the cause of the problem. Furthermore, the standard requires organizations to be proactive and identify actions to eliminate the cause of problems that could potentially occur. The standard requires that organizations have one or more procedures that define these requirements.

The ISO 50001 standard defines a "correction" as an action taken to eliminate a detected nonconformity. As applicable, a root-cause analysis may be conducted to determine the underlying cause of an incident or noncompliance. Based on these findings, *corrective actions* are then taken to ensure that the problem does not happen again.[1] Findings and/or recommendations resulting from EnMS monitoring and the auditing phase provide the basis for identifying and managing preventive/corrective actions. Such preventive/corrective actions can also be designed to promote and maintain sustainability goals and objectives.

An example of a corrective action follows: During the course of an internal audit it was discovered that a piece of equipment not being used was energized and needlessly consuming electricity. A possible correction would be to simply turn the machine off. However, this correction does not prevent the machine from being left on again in the future. Upon conducting root-cause analysis, suppose it was determined that the manufacturing process takes the machine operator away from the on/off switch and into another room to perform the next function, but the operator forgets to return and turn off the machine. A possible corrective action could be to install a timer or electronic sensor that shuts the machine off when it has been idle for a designated period of time. Furthermore, installing these devices on other pieces of equipment as well could be considered a corrective action because the device would prevent these other pieces of equipment from being left on.

In another example, it was discovered that at one organization, the workforce leaves 72% of its desktops turned on overnight and 63% of its desktops turned on over the weekends. It was subsequently determined that if 50% of those desktops were placed into standby (or "sleep") when not in use, the organization could save 2 million kilowatt hours and almost 1,400 tons of carbon dioxide per year. The organization then implemented a program to automatically manage power on its desktops using purchased software. The software allowed the organization's information technology department to manage the power consumption of these desktops by sending them into standby during nonstandard work hours.

Many companies use a corrective action request (CAR) form to address deficiencies. We suggest that, at a minimum, the CAR should document the nonconformance, specify the entity responsible for addressing the problem, and define a course of action for rectifying the nonconformance. It is important to draw a distinction between the terms *nonconformance* and *noncompliance*. The term n*onconformance* refers to deviations from the EnMS requirements. In contrast, the term *noncompliance* refers to deviations from government laws or regulations.

Preventive actions by contrast are intended to address potential nonconformities and other undesirable situations that have not yet occurred, but that you have reason to believe may occur. Emphasis is placed on identifying potential future *nonconformities* or problems, and taking measures to prevent them before they occur. Preventive actions are often identified in the organization's risk management program activities.

Managers sometimes encounter opposition attempting to justify a preventive-action program (when no actual incident or nonconformance has actually occurred); this is because it is often difficult to determine the effectiveness of a resulting initiative that prevents problems from occurring in the first place. The reader should note that a *preventive action* is taken to prevent occurrence, whereas *corrective action* is taken to prevent recurrence.

Corrective and preventive actions should be appropriate to the magnitude of the actual or potential problems and the energy consequences encountered. The organization must ensure that any necessary changes are made to the EnMS documentation.

6.4.2 *How will this benefit us?*

Employees benefit because they receive clear, written instructions for how to deal with actual and potential problems. People are more likely to do the right thing if they know what to do. Another benefit of the corrective action process is that it can be viewed as a "good thing"—unlike traditional noncompliance situations. In other words, an EnMS can actually be used to put a positive "spin" on a less than optimal result; few other scenarios are as forgiving.

6.4.3 *How do we implement this?*

Although the standard does not specifically require that these procedures be documented, the only practical way for anyone to know what to do and for auditors to know what requirements they are auditing against is to write down the requirements and control them according to the organization's control of documents procedure.

Many organizations develop a "corrective action" procedure and a "preventive action" procedure because it tends to keep things simpler. However, ISO 50001 allows these requirements to be captured in a single procedure. This makes sense since the fundamental requirements for corrective action and preventive action are generally quite similar. The only difference is that one occurs after a nonconformity has been detected (corrective action) and the other occurs before a nonconformity has occurred and a designee is implemented to prevent such occurrences (preventive action). The organization should do whatever works the best for its internal size and structure.

An organization may already have documented corrective action and preventive action procedures as part of its quality, environmental, or safety management systems. If so, there is no need to establish redundant procedures. Simply make it clear that the existing procedures apply to energy-related matters and are part of your EnMS. If an organization does not have a corrective action procedure, one will need to be created; the procedure can be as detailed or simple as necessary. Although each organization breaks down the corrective action steps differently, here are the generic steps for managing corrective actions:

1. Collect the nonconformity information and prioritize on the basis of importance and consequence.
2. Apply a root-cause analysis (formal or informal as appropriate) to determine the cause of the nonconformity.
3. Identify the appropriate action, if needed.
4. Implement the appropriate action.
5. Verify to see if the action was properly implemented.
6. Validate to see if the action was effective.

The ISO 50001 standard recognizes that the magnitude of each problem or potential problem may be different. Therefore, organizations should deal with these in a manner commensurate with their associated energy impact or consequence.

6.4.4 What evidence do we need?

Documented procedures that describe how actual and potential nonconformities are dealt with and records of the actions actually taken are needed. This makes it easy for employees to know what to do and for auditors to evaluate how the organization has elected to handle these requirements.

6.5 Reviewing and determining the causes of nonconformities and potential nonconformities (4.6.4a of the ISO standard)

The ISO 50001 requires these procedures to define how the organization will review nonconformities and potential nonconformities and determine their causes.

6.5.1 What does this mean?

The corrective action procedure should define how the organization will review any problems and figure out what caused them. This applies to the preventive action procedure and potential problems as well.

6.5.2 How will this benefit us?

Reviewing nonconformities helps the organization to understand the nature of the problem. Determining what caused a particular problem enables the organization to develop solutions to fix the problem, so it will not occur again.

6.5.3 How do we implement this?

Nonconformities and potential nonconformities should be reviewed by top management during their management reviews. As an input to their review, it would be useful to provide preliminary analysis as to what caused a problem or could cause a potential problem. This is often accomplished using a systematic problem-solving method to isolate the cause. For example, you might start by defining the problem, gathering data, and asking "why" several times to identify the true root-cause of the problem. This analysis helps the energy management team and top management to better understand the issues and ensure appropriate actions are taken to prevent the problem from reoccurring.

Reviewing and determining the causes of potential nonconformities is a little trickier because it is hard to visualize problems that have not yet occurred. The failure modes and effects analysis (FMEA) is a useful methodology for identifying potential nonconformities.* Here are the eight high-level steps for using this methodology:

1. Determine potential failure modes.
2. Determine effects of failure modes.
3. Determine severity of failure effects.
4. Determine causes of failure modes.
5. Determine FMEA occurrence ranking.
6. Determine failure mode controls.
7. Determine prevention/detection of failure modes.
8. Calculate risk priority number (RPN).

6.5.4 What evidence do we need?

Documented records of management reviews provide evidence that management has reviewed nonconformities identified from audits, assessments, and recommendations for improvement. These records of management reviewing preventive actions provide evidence that management has reviewed potential nonconformities as well. Records of causal analysis, root-cause analysis, or any other thought process used by the organization to determine the cause of nonconformities or potential nonconformities also constitute evidence. Causal analysis records of nonconformities are usually associated with an organization's corrective actions procedure implementing tool. This could be a database or electronic tool that helps the organization manage the disposition of corrective actions. Causal analysis records of potential nonconformities are often associated with an organization's risk management, lessons learned, or issues about management procedure.

6.6 Evaluating the need for action (4.6.4b of the ISO standard)

ISO 50001 requires that corrective and preventive action procedures define the organization's requirements for evaluating the need for action to ensure that nonconformities do not occur or reoccur.

---

* For example, see Center for Chemical Process Safety of the American Institute of Chemical Engineers, Guidelines for Hazard Evaluation Procedures, Chapter 4 for more details about this and several other excellent methods. Also see its Chapter 5: Selecting Hazard Evaluation Techniques.

6.6.1 What does this mean?

One of the first steps in deciding what to do about a problem or potential problem is to decide if anything should be done at all. The ISO 50001 standard does not require you to take action on every nonconformity; it only requires that a deliberate thought process be used in deciding which ones require action. The corrective and preventive procedures should define the methodology or process the organization will use in evaluating the need to take action.

6.6.2 How will this benefit us?

Organizations will spend their time addressing only those problems or potential problems that need addressing. Prioritizing nonconformities based on their magnitude and impact enables the organization to focus on the important problems and address the rest as time and resources permit. Many organizations become paralyzed because they try to fix all their problems at the same time. A little thinking up front can save a lot of time on the back end.

6.6.3 How do we implement this?

Develop a thought process or methodology for determining which nonconformities have the potential to impact the organization the most. One example would be to quantify the expected change in energy performance associated with each nonconformity. Then list the nonconformities by order of expected change, from highest to lowest. The organization could then implement the top 20% of the respective corrective actions, or whatever number of actions existing resources permit.

Evaluating the need for preventive action can be a little more difficult because the nonconformity has not yet occurred, and it may be difficult to understand its potential impacts on energy performance. One method is to rank the risk priority numbers calculated from a FMEA and address the top candidates. Another method is to evaluate the likelihood of the occurrence of the potential nonconformity and estimate the consequences based on available information. This can be accomplished subjectively or using a quantitatively rigorous method as needed. This should be routinely performed as part of developing the Energy Plan pursuant to Chapter 4. A suggested format of likelihood definitions is provided in Table 6.1.

Table 6.2 provides a potential set of risk consequences scoring terms that can be used to estimate the impact of a potential nonconformity.

The matrices depicted in Tables 6.1 and 6.2 can be combined to produce a potential nonconformity risk decision guide similar to the one presented in Table 6.3.

Table 6.1 Likelihood Definition Matrix

Level	Probability	Definition
VH	Very high	Likely to occur often. Likelihood of occurrence is estimated to be greater than 0.10 per operational opportunity.
H	High	Expected to occur sometime in the life of the energy program. Likelihood of occurrence is estimated to be between 0.01 and 0.10 per operational opportunity.
M	Moderate	Likely to occur sometime in the life of the energy program. Likelihood of occurrence is estimated to be between 0.001 and 0.01 per operational opportunity.
L	Low	Unlikely but possible to occur. Likelihood of occurrence is estimated to be between 0.000001 and 0.001 per operational opportunity.
VL	Very low	Likelihood of occurrence is estimated to be less than 0.000001 per operational opportunity.

Table 6.2 Risk-consequences Scoring Term

Level	Matrix
1	**Negligible** impact on system capabilities or technical performance of the EnMS.
2	**Minor** problems with the EnMS performance or capabilities that can be easily fixed.
3	**Major** problems with the EnMS capabilities or performance, but EnMS is still functional.
4	**Severe** degradation of EnMS performance or capabilities; major impact on EnMS functionality.
5	**Extreme** degradation of EnMS performance with catastrophic consequences.

This kind of guide is a tool that helps management apply a risk- and consequence-graded approach to decision making, using the following steps:

1. Define applicable categories of risk and consequences that make sense to your organization's energy system decision.
2. Discuss the meaning of each matrix cell and decide whether it merits "accept" or "reject." For example, in Table 6.3, we decided to reject very high consequences no matter how low their likelihood, and to accept all very low likelihoods except for what we called *catastrophic consequences*.
3. The final step is to identify where each relevant decision option maps onto management's own decision guide—and seek consensus.

Chapter six: Checking performance 111

Table 6.3 Potential Nonconformity (NC) Risk Decision Guide for Five Sample Assessments

Five potential nonconformities rated by risk		Decision for using/not using in five cases				
	1	2	3	4	5
VH	Accept	Reject	Reject	Reject	Reject
H	Accept	Accept	Reject	Reject	Reject
M	Accept	Accept	Reject	Reject	Reject
L	Accept	Accept	Accept	Reject	Reject
VL	Accept	Accept	Accept	Accept	Reject
Risk level	Negligible	Minor	Major	Severe	Catastrophic

Note: VH, very high; H, high; M, moderate; L, low; VL, very low.

In some cases, there may be reasons to reconsider the indicated decision guidance.

6.6.4 *What evidence do we need?*

The organization's corrective action procedure should include a step for evaluating the need for corrective action and a definition of how the organization will conduct these evaluations. EnMS records should include identification of the nonconformities, any corrections you made, and your analysis of the root causes. Also retain documented records of your prioritization and the decision about which corrective actions to implement.

6.7 *Implementing corrective and preventive actions (4.6.4c of the ISO standard)*

The standard requires that corrective and preventive action procedures define the requirements for determining and implementing the appropriate action needed.

6.7.1 *What does this mean?*

The organization's corrective action procedure should provide direction on how to decide what actions need to be taken to fix an immediate problem and prevent its reoccurrence. The procedure should very clearly distinguish between remedial actions designed to fix a particular problem (corrections) and actions designed to address the root cause of the problem (corrective actions).

The procedure should define who is responsible for addressing corrective actions. Some issues can be managed within the control of a single department, and the authority to implement may lie with a single person. Other solutions may require cooperation from individuals throughout the organization. In this instance, a corrective action team is preferred rather than a single individual. People are more willing to implement changes when they have had a hand in determining needed changes through problem solving.

6.7.2 How will this benefit us?

Implementing actions to eliminate nonconformities leads to improved performance of the EnMS and ultimately improved energy performance.

6.7.3 How do we implement this?

The first step is to fix the specific nonconformity. This is normally a temporary measure, a correction designed to immediately address the problem for the short term. An intermediate next step would be to look at other similar situations to see if the problem is more widespread than previously thought. The next step is to study the root cause of the nonconformity and design actions that ensure the same problems do not happen again in the future. Take some time to consider whether there is a possibility of your improvement actions introducing new problems into your EnMS. For complicated problems or those that cross-cut the organization, it may be helpful to use a team approach to determining the appropriate action needed.

6.7.4 What evidence do we need?

Maintain the records of corrective and preventive actions you need to trace nonconformities and risk-based preventive action opportunities from identification through implementing actions and validation of effectiveness. The following section has a more comprehensive list of records that should be maintained.

6.8 Maintaining records of corrective and preventive actions (4.6.4d of the ISO standard)

The ISO 50001 standard requires that corrective and preventive action procedures define the requirements for maintaining records of corrective and preventive actions.

6.8.1 What does this mean?

This means that the organization's corrective and preventive action procedures should specify what records should be generated and how the organization will maintain the records of corrective and preventive actions.

Chapter six: Checking performance 113

6.8.2   How will this benefit us?

The organization benefits by an improved EnMS that should save energy. The idea of continual improvement is that the organization has a built-in mechanism for identifying problems, fixing them, and ensuring that the problems are solved once and for all. Maintaining records of this whole process allows the organization to assess how well it is accomplishing this.

6.8.3   How do we implement this?

Decide what records you will need in order to trace problems from identification to resolution and define these in your corrective and preventive action procedures. If the organization is ISO 9001- or 14001-certified, it already has these procedures and they define the needed records.

Create the records and maintain them for as long as they are useful. Here are related activities for which you will need records as part of your corrective and preventive action procedures:

- Identifying the nonconformities or potential nonconformities
- Reviewing nonconformities to understand the problem and severity
- Determining the root-cause of nonconformities
- Evaluating the need for action
- Determining what actions are needed
- Verifying that action have been implemented
- Validating that actions were effective

6.8.4   What evidence do we need?

Defined requirements for corrective and preventive action activities and the related records are needed.

6.9   Reviewing the effectiveness of the corrective or preventive action taken (4.6.4e of the ISO standard)

The ISO 50001 standard requires that corrective and preventive action procedures define the requirements for reviewing the effectiveness of the corrective or preventive action taken.

6.9.1   What does this mean?

The proper management should verify that corrective and preventive actions were properly implemented according to defined actions. After the actions have been in place and have had an opportunity to realize their intended impact, management should look at the corrective action again to validate that the actions were effective and led to the desired results.

6.9.2 How will this benefit us?

Ensuring that problems do not repeat themselves or even happen at all leads to improved energy performance.

6.9.3 How do we implement this?

After corrective actions have been implemented, the responsible management should look for evidence of implementation according to the actions specified in the corrective actions records. After a reasonable period of time has passed in which the actions have had a chance to take effect, the responsible management should validate to see if the actions were effective. Have they prevented reoccurrence of the problem? Have they created new problems? Are further actions required?

6.9.4 What evidence do we need?

Any documented records showing that management has verified and validated corrective and preventive actions are needed.

6.10 Control of records (4.6.5 of the ISO standard)

The ISO 50001 standard requires an organization to establish and maintain records as necessary to demonstrate the energy performance results achieved as well as conformity to the requirements of its EnMS and the ISO 50001 standard.

6.10.1 What does this mean?

In the ISO 50001 standard, the term "records" generally refers to a written account of knowledge, facts, or events related to the organization's EnMS and energy performance. Examples of records could include reports stating energy performance results, the analysis used to identify corrective actions, and management review agendas and meeting minutes that demonstrate management reviews have taken place. Records help the organization demonstrate to its top management and ISO 50001 auditors of its energy performance that it is adhering to the requirements of its EnMS and is in compliance with the ISO 50001 standard. Table 6.4 provides a potential list of records that need to be maintained.

6.10.2   How will this benefit us?

Written records foster a common understanding among multiple people. They make information accessible to people who may not have personally prepared a particular performance report or were not present at a particular event, such as a management review. Information contained in written records tends to be more accessible to others and tends not to change over time as often happens with information residing in people's heads. Written records also provide documented evidence for the ISO 50001 registration process.

6.10.3   How do we implement this?

The organization should first decide how they will identify, store, retrieve, and control their energy records. Although the ISO 50001 standard does not require organizations to have a documented procedure for controlling records, other ISO standards, including ISO 9001, do so with good reason. Detailed requirements are nearly impossible for everyone to memorize in their heads, and documenting them benefits the organization in the ways described earlier. Next, the organization should identify the different types of EnMS records they use. Table 6.4 provides a good summary of the types of energy-related records an organization will use. Finally, in a manner suitable to your organization, maintain the records, so they

Table 6.4 List of Potential Records

- Energy review including the methodology and criteria used to develop it
- Opportunities for improving energy performance
- Energy baseline
- Energy performance indicators
- Competence, training, and awareness of employees
- Decision on whether to externally communicate its EnMS and energy performance criteria and results
- Design activity results
- Key operational characteristics monitoring and measurement results
- Calibration records
- Evaluation results of energy use related to legal and other compliance
- Internal audit schedules and results
- Corrective and preventive action records
- Management review activities

can be retrieved when necessary. Small organizations may choose to store their records in a central location. Larger, decentralized organizations may not specify where the records are stored, so long as they can be retrieved within a designated timeframe. The ISO 50001 standard further requires that organizations ensure that their records remain legible, identifiable, and traceable to the relevant activity, product, or service.

Endnote

1. Pojasek, R.B., "Introducing ISO 14001 III," *Environmental Quality Management*, Autumn 2007.

chapter seven

Management review

This chapter describes the ISO 50001 requirements underlying management review of an ISO 50001 Energy Management System (EnMS), which is described in ISO 50001, Section 4.7. We begin by describing the requirement that requires top management to review the organization's EnMS at planned intervals.

7.1 Summary of requirements

ISO 50001 requires that top management review the organization's energy management system at planned intervals to ensure its continuing suitability, adequacy, and effectiveness.

7.1.1 What does this mean?

This means that top management regularly and systematically reviews the organization's energy-related information, evaluates the information, allocates resources, and directs that improvement actions be undertaken when necessary. "Planned intervals" means that the reviews are conducted according to the organization's own schedule. The organization's energy-related information is reviewed to ensure that the EnMS is still appropriate for the organization and it meets the organization's own requirements for achieving expected energy performance results.

7.1.2 How will this benefit us?

This involves top management in the improvement of the EnMS. The clear direction and appropriate resource from an informed management team will result in a more effective EnMS and improved energy performance. Their assessment of whether plans have been executed and targets achieved will result in the necessary actions for improvement.

7.1.3 How do we implement this?

First, top management must review the EnMS process, procedures, tools, and information to determine all the right pieces of the system are present in order to manage energy performance. For the right pieces that are in place, management must next determine if they are appropriate for sustaining the organization's current energy performance. For the processes and procedures not in place, management must direct that they be developed and implemented. Lastly, top management must assess the extent to which planned energy activities are completed and expected energy performance is achieved.

7.1.4 What evidence do we need?

The organization should create and maintain management review records. Agendas, information used during the review, and meeting minutes are all helpful. Meeting minutes are especially useful for documenting decisions and action items and accessing them at a later time.

7.2 Input to management review (4.7.1 of the ISO standard)

The standard requires that management reviews include information about various facets of the organization and its energy management system.

7.2.1 What does this mean?

An effective management review includes reviewing and discussing certain information about the organization. Information from EnMS audits, energy performance, and *energy performance indicators* must be analyzed to determine the extent to which the energy policy, objectives, and targets are being met, and identify opportunities for improvement. Examining system performance is done under the backdrop of legal and other requirements, as well as the organization's overall goals and objectives.

7.2.2 How will this benefit us?

Historical information reveals that management reviews tend to be more effective and useful when certain topics, or inputs, are discussed. Understanding how their systems work and the resulting performance enables top management to make decisions that will improve the overall organization and its energy efficiency.

7.2.3 How do we implement this?

Scheduling adequate time and resources for discussing the relevant topics during management review meetings is critical. Many organizations prepare a meeting agenda that specifies the topic to be addressed during the management review. This can be accomplished in a single meeting or a series of meetings, depending on the needs of the organization. Some organizations conduct management reviews weekly, while others do so on an annual basis. Perhaps somewhere in between would be appropriate for your organization. The important thing is to determine what would work best for the organization and stick to it.

The inputs (Items [a] through [i]) required by the standard, which are discussed in the following sections, are necessary but may not be sufficient for every organization. Management reviews should include any information that is required for management to determine the effectiveness of the EnMS and act accordingly on their determination. Also keep in mind that an organization probably already has several management review venues. There is no requirement to review the organization's EnMS separately from other quality (ISO 9001), environmental (ISO 14001), or other management systems. Feel free to integrate the energy review with other existing reviews. Because your EnMS is likely integrated with your other systems, there is a good chance that many of your processes and procedures are already being reviewed.

7.2.4 What evidence do we need?

Many organizations document that their management reviews include the required inputs using meeting agendas and the associated information that was reviewed. One could reference reports, e-mails, or any other sources of information relied on during the management review. One can even videotape management reviews. During an ISO 50001 audit, the auditor will be looking for documented evidence that the required topics (inputs) were discussed during management reviews at least as often as you have prescribed for yourself.

7.3 Actions from previous management reviews (4.7.1a of the ISO standard)

A major premise of management reviews is that upon assessing overall performance management identifies actions to maintain or improve that performance. To ensure these actions are carried out, their status must be reviewed during subsequent management reviews. The outputs of previous management reviews will include action items that become inputs to the current management review. A review of these follow-up actions should include whether the action is open or closed. If it is open, what is its status relative to the assigned deadline, what progress has been made, and what is the prognosis for when it will be completed? If it is closed, what was accomplished, what evidence is there to verify its completion, and was it effective?

7.4 Energy policy (4.7.1b of the ISO standard)

Because the energy policy is established to provide a focus for directing the organization's energy performance improvement activities, top management must determine whether the energy policy can be used to help define the desired results and assist the organization in applying its resources to achieve these results. The energy policy provides a framework for establishing and reviewing energy objectives. The energy policy should state the organization's commitment for achieving energy performance and provide a framework for setting and reviewing energy objectives and targets. Although one would not expect an organization's commitment to substantially change very frequently, it may occasionally happen. The energy policy should be regularly reviewed and updated as appropriate.

7.5 Energy performance and energy performance indicators (4.7.1c of the ISO standard)

Top management must assess the organization's energy performance by reviewing the energy performance indicators and comparing them to the energy baseline on a regular basis. Management must ensure that the energy baseline is adjusted when the energy performance indicators no longer reflect organizational energy use or there have been major changes to the EnMS.

7.6 Legal compliance and requirement changes (4.7.1d of the ISO standard)

Under ISO 50001, top management must review an evaluation of legal compliance and changes in legal and other requirements to which the organization subscribes.

The reason for doing so is that the energy policy, objectives, targets, and action plans must be taken into account. Top management must have a clear understanding of how these requirements interact with the organization's EnMS and impact energy performance.

7.7 Energy objectives and targets (4.7.1e of the ISO standard)

Meeting the organization's commitment to energy performance is accomplished by setting energy objectives that are aligned with the energy policy. Information should be reviewed to determine if these specific outcomes and achievements are being accomplished. The key to accomplishing each energy objective is to set one or more energy targets, such that if the targets are met, the objective is achieved. Energy targets should be quantifiable, applicable to the organization, and directly relevant in meeting one or more energy objectives. Information should be reviewed to establish whether the energy targets are being achieved. Improvement opportunities for meeting energy objectives and targets include

- Improving process efficiency relative to energy use
- Revising processes and procedures to enable energy performance improvement activities
- Revising energy targets to better align with energy objectives
- Restating energy objectives to more accurately reflect the organization's energy policy commitments

7.8 Energy management system audit results (4.7.1f of the ISO standard)

Reviewing EnMS audit results will help management determine if their system is being implemented according to requirements and if it is robust enough to fulfill the commitments in the energy policy. Understanding system problems in the context of the energy policy helps management focus its limited resources and prioritize improvement actions.

Since the authority and resources for establishing the organization's energy audit program come from top management, it is appropriate that they review the overall achievements of the audit program to assess whether its objectives have been met and to identify opportunities for improvement. An audit program that is capable of detecting noteworthy and relevant problems adds value to the organization by expanding its energy performance capabilities.

7.9 Corrective and preventive actions (4.7.1g of the ISO standard)

Corrective action information should be reviewed to determine if actions are being taken to eliminate the causes of detected nonconformities. A review of these corrective actions should include whether the action is open or closed. If it is open, what is its status relative to the assigned deadline, what progress has been made, and what is the prognosis for when it will be completed? If it is closed, what was accomplished, what evidence is there to verify its completion, and was it effective? This is an excellent opportunity for management to review solutions that require new capabilities, additional resources, or new energy system processes.

Preventive action information should be reviewed to determine if actions are being taken to eliminate the causes of potential nonconformities. It follows then that a review of preventive actions include consideration of whether any nonconformities could have been prevented, and if so, why were they not? While no new nonconformities may seem like good news for the risk management and preventive action processes, it could also mean that the internal audit program is not effectively identifying existing problems.

It is important to remember that the outcome of your neverending efforts to balance these competing forces is not necessarily intended to be punitive but rather introspective. The primary objective is to continually improve the organization's EnMS and ultimately its energy performance.

A review of existing preventive actions should include whether the action is open or closed. If it is open, what is its status relative to the assigned deadline, what progress has been made, and what is the prognosis for when it will be completed? If it is closed, what was accomplished, what evidence is there to verify its completion, and was it effective? High-performing management teams base their reviews on objective evidence rather than assertion.

Many organizations will discover that they already consider preventive actions as part of their enterprise risk management activities. We recommend this as a best practice in order to fully integrate the EnMS with the organization's other management systems.

7.10 Projected energy performance (4.7.1h of the ISO standard)

Current energy performance information should be reviewed to determine projected energy performance for the following period, as appropriate. This allows management to compare projected performance with the targets and make any adjustments necessary in advance to meet the targets. Potential adjustments might include acquiring the capability to measure energy use, modifying system processes, communication, or training.

7.11 Recommendations for improvement (4.7.1i of the ISO standard)

Recommendations for improvement should be reviewed so that management can identify, prioritize, and authorize improvement activities. These recommendations may originate from anywhere in the organization and make their way into the management review venue, or be identified during the course of the management review by way of reviewing the inputs discussed earlier. Many improvement activities require new or improved energy system processes, new organizational capabilities, or additional resources; top management commitment is essential for implementing these changes.

7.12 Output from management review (4.7.2 of the ISO standard)

The standard requires that outputs from the management review include any decisions or actions related to specific aspects of the organization's EnMS and energy performance.

7.12.1 What does this mean?

For management reviews to be truly useful to the organization, something must be done with the information presented to top management. The information must be used to make decisions and take actions that will ultimately improve the organization's EnMS and energy performance.

7.12.2 How will this benefit us?

Decisions and actions will result in an improved EnMS and improved energy performance.

7.12.3 How do we implement this?

Each management review should include not only time to review information but to also discuss the information reviewed, determine its meaning, assess whether the results are acceptable, and take action when they are not. High-performing management teams base their reviews on objective evidence rather than assertion. (See items [a] through [i] of this section as explained below.)

7.12.4 What evidence do we need?

Top management's energy-related decisions and actions should be documented in the management review meeting minutes in an action item management tool or by whatever means the organization uses to memorialize its top management activities.

7.13 Changes in the energy performance of the organization (4.7.2a of the ISO standard)

The standard requires that the output of management review includes decisions and actions related to changes in the energy performance of the organization.

Top management reviews the organization's energy performance to decide if it is remaining constant, improving, or getting worse. Their decisions are documented in the management review meeting minutes. If energy performance is getting worse, they authorize and document the actions they determine appropriate for improving their energy performance.

7.14 Changes to the energy policy (4.7.2b of the ISO standard)

The standard requires the output of management review to include decisions and actions related to changes in the organization's energy policy.

The purpose of reviewing the organization's energy policy is to determine if it still reflects the organization's commitment for achieving energy performance improvement and to provide a suitable framework for doing so. If top management determines that it does, then simply document the decision. If it is determined that changes are required, document the decision and appropriate changes to be made.

7.15 Changes to the EnPIs (4.7.2c of the ISO standard)

The standard requires that the output of management review include decisions and actions related to changes in the organization's energy performance indicators.

Top management should decide whether the organization's energy performance indicators are appropriate for monitoring and measuring energy performance. If so, document the decision in the meeting minutes. If the performance indicators no longer reflect organizational energy use, document what actions are to be taken based on the organization's methodology for updating the energy performance indicators. Namely, the performance indicators will have to be modified and the energy baseline will need to be adjusted.

7.16 Continual improvement of the energy management system and its implementation (4.7.2d of the ISO standard)

The standard requires that management review output includes decisions and actions related to changes to objectives, targets, or other elements of the EnMS, consistent with the organization's commitment to continual improvement.

Continual improvement of the organization's energy performance and the EnMS depends on a well-defined set of objectives and targets designed to achieve the energy policy. Accordingly, top management should take action to ensure the EnMS is capable of implementing the objectives and targets.

7.17 *Allocation of resources (4.7.2e of the ISO standard)*

The standard requires that the output from the management review include any decisions and actions related to the allocation of resources. Improving energy performance and the EnMS typically requires personnel to do the work, money to implement solutions, and the physical space from which to work. These resources tend to be limited and top management is responsible for prioritizing improvement actions and providing the necessary resources for implementation.

This benefits the organization because improvement actions that are appropriately funded and staffed have a better chance of being successfully implemented. Top management makes decisions, documents the decisions, and the organization uses its internal processes for allocating resources and completes the actions per management direction.

High-level decisions and actions related to resource allocations should be documented in the management review minutes. Annual budgets and plans that contain allocations are also a good source of more detailed evidence. ISO auditors also find evidence of these decisions indirectly by observing the organization's energy activities. For example, if an ISO auditor observes several members of an EnMS team performing a procurement process improvement exercise designed to include energy performance as part of the supplier evaluation, the auditor can see that actual personnel and dollars are being used to improve the process.

Bibliography

Hoyle, David, *ISO 9000 Quality Systems Handbook*, Fifth Edition.
ISO 9001:2008.
ISO 50001:2011.
ISO 9000—definitions.
ISO 190011—auditing.
FMEA source.
Risk matrices sources.

Appendix A: Energy consumption, generation, sustainability, and energy systems

A.1 Energy consumption

Modern society converts energy from forms that are less desirable and typically less dense to forms that are more desirable. For example we convert grains and grazing grasses to meat (cows and other livestock), oil to gasoline, which powers vehicles, and coal to electricity, which powers manufacturing plants. The primitive human, who had yet to discover the fascination of fire, had access only to the food he ate, which was used by his body as energy to power the activities needed for survival. Soon man developed the hunting skills that provided more energy per gram of food consumed. About 100,000 years ago, primitive humans developed better methods of acquiring food and also of burning wood for both heating and cooking. Energy consumption dramatically rose again as humans learned to harness the power of animals to aid in growing crops. Energy consumption again rose in a dramatic leap as our ancestors invented devices to tap the power of wind and water. This eventually led to burning a much denser type of energy resource—coal. The industrialization age was driven by the invention of the steam engine, which again resulted in a dramatic increase in energy generation. The steam engine allowed us to unlock the Earth's vast storage deposits of solar energy (coal, gas, and oil). Prior to industrialization, these increases in energy consumption had been slow and gradual. Once industrialization came to center stage, the rate of energy consumption increased exponentially over a period of just a few generations.

A.1.1 Energy consumption data

Figure A.1 illustrates global energy consumption between 1850 and 2000, compared to the increase in global population over the same period. As

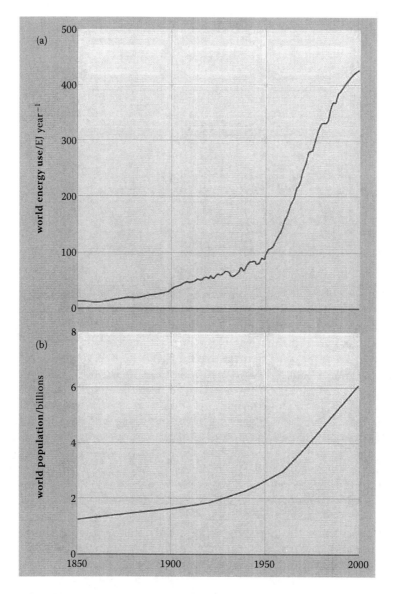

Figure A.1 The top plot shows global energy consumption between 1850 and 2000. The bottom plot graphs population over time. These two figures illustrate that energy consumption is increasing faster than population.

Appendix A

the figure demonstrates, energy consumption is growing faster than population.

Energy consumption is ten times greater than it was in the beginning of the 20th century. Major forms of energy over the 20th century were nonrenewable sources:

- Coal
- Oil
- Natural gas
- Nuclear energy

Oil, coal, and natural gas are also known as fossil fuels. Two principal problems with nonrenewable energy sources are limited quantity and environment pollution, particularly the generation of large quantities of CO_2, a greenhouse gas. Many scientists believe the increase in greenhouse gases is the principal reason for the increase in global temperature in recent decades. Nuclear power plants do not produce appreciable amounts of greenhouse gases, but their waste remains radioactive for generations and needs to be isolated from the biosphere. The principle renewable energy sources are:

- Wind energy
- Sun energy
- Bioenergy
- Water energy

As the global population increases, particularly in developing countries, and their citizens demand modern standards of living, global energy use will continue to increase, with developing nations accounting for a growing share of total world demand (See Figure A.2).

Projections of total world energy consumption through the year 2300 are shown in Figure A.3, where it is indicated that consumption is expected to increase rapidly. Eventually, it will begin to level out around the year 2175.

A.1.2 Energy consumption by industrial sector

Figure A.4 shows projections of total world energy consumption. The plot depicts rapidly increasing demand, particularly among the developing world, through most of the 21st century.

A.1.3 Energy consumption by fuel type

Figure A.5 shows projected world energy consumption by fuel type. This projection indicates a much faster increase in fossil fuel use than in either

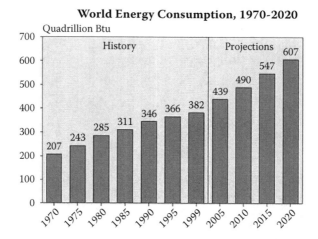

Figure A.2 Projected increase in world energy consumption through the year 2020. The figure shows a consistent increase in energy consumption through the year 2020.

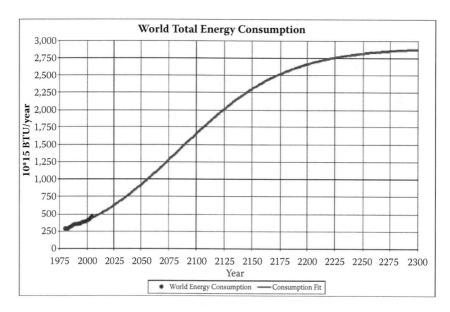

Figure A.3 Projections of total world energy consumption through the year 2300.

Appendix A

Energy consumption by sector
Total U.S. energy consumption by end-use sector, 2007

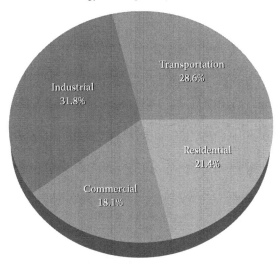

Figure A.4 Shows by industrial sector total world energy consumption.

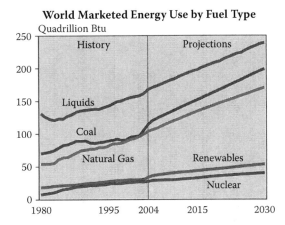

Figure A.5 Projected world energy use by fuel type (From the International Energy Agency.)

renewable or nuclear energy. Peak oil advocates question whether such a sharp increase in fossil fuel use is possible or can be sustained.

A.1.4 Relationship between GDP and energy consumption

Figure A.6 shows the GDP/capita for various countries versus how much energy they consume (kW/capita). While it appears that there is a relationship to the U.S. GDP and its energy consumption, this relationship does not hold for all nations. While Japan and Russia have the same energy usage, Japan has 10 times the GDP of Russia.

Figure A.7 shows which countries use more than their share of the world's energy as compared to their GDP per capita. For instance, the United States has approximately 5% of the world's population but consumes 23% of its total energy. The Russian Federation has 2% of the world's population and consumes 7% of the world's energy. Meanwhile, China holds 20% of the world's population and consumes only 17% of the world's energy.

A.2 Energy production

As depicted in Figure A.8, the majority of energy produced around the world comes from fossil fuels. While renewable energy sources do not

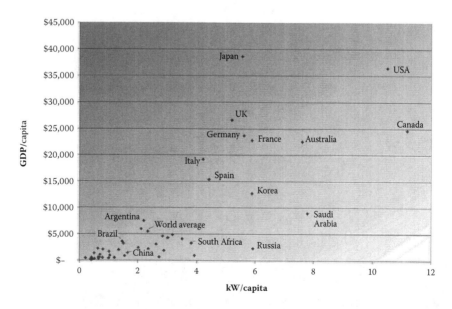

Figure A.6 Relationship between GDP/capita and energy consumption/capita.

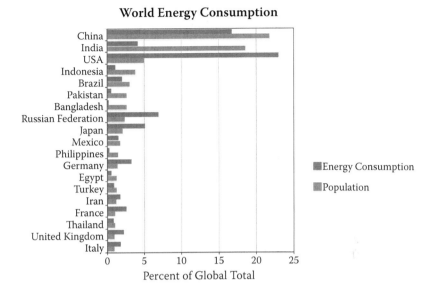

Figure A.7 National consumption of energy versus the nation's population base. The United States clearly ranks as the most energy-intensive country per capita. Meanwhile the Philippines and Bangladesh have among the lowest consumption per capita.

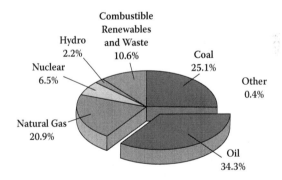

Figure A.8 Breakdown of energy production by source. (From Key World Energy Statistics, International Energy Agency.)

pollute the environment in the same amount as nonrenewable sources, they nevertheless are not completely clean or without their own set of environmental problems. For instance, burning biomass results in CO_2 emissions similar to that of fossil fuels. The principal problems with renewable energy sources are still the price and relatively small contribution to global energy production. However, the cost of renewable energy sources is generally declining; while it has significant potential, this technology will have limited use in the near future.

Switching from one energy resource to another involves more than simply developing a new technology. It also involves altering the systems that produce, process, and transport these resources. Primary commercial energy fuels like coal, natural gas, and uranium, for example, are mined, processed, refined, and transported through complex systems that represent billions of dollars in infrastructure investments and complicated technological interconnections. Typically, energy production or refining facilities operate for 30 to 60 years or more; we cannot simply switch to different energy sources or technology mixes overnight. Retiring these existing systems prematurely to replace them with something better is a very expensive proposition even if the new facilities are more efficient.

A.2.1 Renewable energy

Renewables already provide almost double the energy production of nuclear power. However, most of the renewable energy comes from burning biomass (wood, crop residue, dung, etc.) and hydropower. Of the remaining 0.5%, over 80% of that comes from geothermal—almost all based on tapping surface volcanic and hydrothermal heat (essentially nothing to date from hot dry rocks). However, geothermal power has large growth potential. What has been referred to as "technosolar" (wind, solar thermal, solar PV, wave) constitutes just over 0.1% of energy production.

A.2.2 The smart grid

One of the most promising technologies for conserving energy involves the *smart grid* (SG). Twentieth-century power grids "broadcast" power from a few central generators to a large number of consumers. In contrast, the smart grid (SG) is capable of optimally routing power so as to respond to a wide range of dynamic conditions, and to charge a premium to those that use energy during peak hours. While there are some proven smart grid technologies in use, the term *smart grid* refers to a set of related technologies on which a specification is generally agreed rather than a name for a specific technology.

A number of governments around the world, including the United States, are currently studying or actively designing SG technology. The SG uses digital technology to deliver electricity to consumers. Two-way communications are combined with net metering technology to monitor grid conditions and measure electrical generation and consumption needs along various parts of the grid. Intelligent monitoring devices keep track of all electricity

Appendix A

flowing across the grid. By monitoring grid usage, automated devices can be used to dynamically respond to changes in grid usage conditions.

Events that may affect grid conditions can include a plant being removed from the grid or perhaps clouds blocking the sun and reducing the amount of solar power to the grid, or a hot day requiring increased use of air conditioning by consumers. The grid would encourage customers to change their usage practices to take advantage of reduced energy costs during periods of low demand and increased prices during peak demand, a process referred to as *peak curtailment*. It is thought that consumers and businesses will tend to consume less during high demand periods if it is possible for consumers and consumer devices to be aware of the high price premium for using electricity at peak periods.

The total load over the power grid can vary significantly. The overall load is not stable and may vary rapidly. Consider the incremental load if a popular television program begins and millions of viewers turn on their televisions, which instantly increases demand. To respond to such a rapid increase in power demand, some spare generators are brought on line. A smart grid could help compensate for such rapid changes.

Consumer electronics devices consume over half the power used in a typical U.S. home. Possessing the ability to shut down or hibernate such devices when they are not actually being used presents a major opportunity for reducing energy consumption; however, this means the utility company can monitor whether you are using your computer or TV or not, and if so, the electric company could decide your computer is not being used and shut it off. Appliances that could aide in the effort to reduce peak demand include air conditioning units, electric water heaters, pool pumps and other high wattage devices. For example, water heaters might be turned off in the middle of the night and pool pumps run during periods of low demand. Because consumer activity can be monitored, privacy advocates warn that this technology may open up a disturbing potential for abuse.

Such a system could lead to significant increased efficiency, reduce electrical costs and greenhouse emissions, and enhance emergency grid resilience during an emergency. In 2009, it was learned that spies had infiltrated the U.S. power grids, perhaps as a means of practicing for an attack on the grid at a later time.[1] An SG can be designed to better absorb or ward off such attacks. By monitoring grid usage, automated devices can be used to dynamically respond to changes in grid usage conditions. In an emergency situation, a localized section of the SG would be able to function independently of other parts of the grid and would have the capability to ration whatever power is available to critical needs, such as communications, hospitals, and police and fire response. It also opens up the potential for entirely new services or improvements on existing ones, such as fire monitoring and alarms that can shut off power or make phone calls to emergency services.

When power is least expensive, the user can allow the smart grid to turn on selected appliances such as washing machines or factory processes that can run at arbitrary hours when electrical supply is high and demand is low. Conversely, at peak times it could turn off selected appliances to reduce demand.

Another key advantage is that it would make it much easier to integrate renewable electricity such as solar and wind into the grid. Businesses and homes are beginning to generate more wind and solar electricity, enabling them to sell surplus energy back to their utilities. However, the existing grid is not designed to efficiently manage such activities. Through bi-directional metering, the SG could more efficiently manage such "traffic" and compensate local producers of power. One United States Department of Energy study calculated that internal modernization of U.S. grids with smart grid capabilities would save between 46 and 117 billion dollars over the next 20 years.[2] Some cities are already constructing SGs. In the United States, Austin, Texas, has been building its smart grid since 2003, when its utility first replaced 1/3 of its manual meters. Boulder, Colorado, completed the first phase of its smart grid project in 2008. Hydro One, in Ontario, Canada, is constructing a large-scale SG initiative. The City of Mannheim in Germany is using real time communications in its Model City Mannheim "MoMa" project.

According to the U.S. Department of Energy's Modern Grid Initiative report, a modern SG should:[3]

1. Motivate consumers to actively participate in operations of the grid
2. Be able to heal itself
3. Resist attack
4. Provide higher quality power that will save money wasted from outages
5. Enable higher penetration of intermittent power generation sources
6. Accommodate all generation and storage options
7. Run more efficiently
8. Enable electricity markets to flourish

A.3 Sustainability

Concerns over the security of energy sources and environmental concerns related to global warming are expected to move the world's energy consumption away from fossil fuels to more sustainable sources. The concept of Peak Oil (see Appendix C) shows that about half of the available petroleum resources have been produced and predicts a decrease of production in the very near future. Many experts predict that oil will be largely exhausted by the year 2022, leaving Americans with the question of how and what will be used as a substitution for this finite energy

Appendix A 139

source. This prediction has led many policymakers to call for an immediate and massive effort to develop more sustainable energy resources. The antithesis of sustainability is a disregard for limits, which has been referred to as the *Easter Island Effect* (see the author's companion text, *Global Environmental Policy: Concepts, Principles, and Practice*), which is the concept of being unable to develop sustainability, resulting in the depletion of natural resources.

A number of nations are taking action as a result of the Kyoto Protocol (global warming) to move towards more sustainable and less polluting energy sources. For instance, the European Commission has proposed that the energy policy of the European Union should set a binding target of increasing the level of renewable energy in the EU's overall mix from less than 7% in 2007 to 20% by 2020. Many critics believe this commitment is overly optimistic.

A.4   A survey of energy sources

This section surveys some of the principal standard energy production technologies as well as emerging technologies and strategies.

A.4.1   Fossil fuels

Fossil fuel energy involves burning hydrocarbon fuels (coal, oil, gas). Heat from burning fossil fuel is used either directly for (1) space and process heating or (2) converted to mechanical energy for vehicles, industrial processes, or electrical power generation.

Table A.1 compares the advantages and disadvantages of fossil fuel plants. Large amounts of greenhouse gas emissions are generated from burning fossil fuels. Currently governments subsidize fossil fuels to the tune of about $500 billion a year. The following sections describe new coal and gas-fired plant technology.

A.4.2   Supercritical coal-fired generation

Coal-fired generation accounts for a greater share of U.S. electrical power generation than any other fuel.[4] Furthermore, the EIA projects those coal-fired power plants will account for the greatest share of added capacity through 2030—more than natural gas, nuclear, or renewable generation options. Supercritical technologies are increasingly common in new coal-fired plants. Supercritical plants operate at higher temperatures and pressures than most existing coal-fired plants (beyond water's "critical point," where boiling no longer occurs and no clear phase change occurs between steam and liquid water). Operating at higher temperatures and pressures allows this coal-fired alternative to function at a higher thermal

Table A.1 Advantages and Disadvantages of Fossil Fuel Energy

Advantages	Disadvantages
• Fossil fuel technology and infrastructure already exist. • Fossil fuels are currently more economical, and practical for decentralized energy use. • Petroleum energy density in terms of volume (cubic space) and mass (weight) is very high compared to most alternative energy sources. • Oil in the form of diesel and gasoline currently provides the only practical and large-scale source of energy for vehicular transportation.	• Fossil fuels are nonrenewable, unsustainable resources, which will eventually decline in production and become exhausted, with severe consequences to societies that are dependent on them. • Extracting fossil fuels is becoming ever more difficult and expensive as we consume the most easily accessible deposits. • Petroleum-powered vehicles are very inefficient. Only about 30% of the energy from the fuel is converted into mechanical energy. • Combustion of fossil fuels leads to the release of contaminants into the atmosphere. A typical coal plant produces in one year[16]: • 3,700,000 tons of CO_2, regarded by many to be the leading cause of global warming • 10,000 tons of sulfur dioxide, the leading cause of acid rain • 500 tons of small airborne particles which can result in chronic bronchitis and aggravated asthma • 10,200 tons of nitrogen oxides, which can lead to formation of ozone and smog, which can cause respiratory illness • 720 tons of carbon monoxide, which can cause headaches and stress on people with heart disease • 220 tons of hydrocarbons, toxic volatile organic compounds, which form ozone • 225 pounds of arsenic, which can cause cancer • 114 pounds (52 kg) of lead, 4 pounds (1.8 kg) of cadmium, other toxic heavy metals. • Dependence on fossil fuels from unstable regions creates energy security risks. Oil dependence has led to war, funding of radical terrorists, and social unrest. • Extraction of fossil fuels results in extensive environmental degradation, such as from coal strip mining. • Burning of fossil fuels in vehicles, buildings, and power plants contributes to urban heat islands.

Appendix A 141

efficiency than most existing coal-fired power plants. While supercritical plants are more expensive to construct, they consume less fuel for a given output, reducing environmental impacts.

In a supercritical coal-fired power plant, burning coal heats pressurized water. As the supercritical steam/water mixture moves through plant pipes to a turbine generator, the pressure drops and the mixture flashes to steam. The heated steam expands across the turbine stages, which then spin and turn the generator to produce electricity. After passing through the turbine, any remaining steam is condensed back to water in the plant's condenser.

A.4.3   Natural gas combined-cycle generation

Natural gas fueled 22% of electric generation in the United States in 2007; this accounted for the second greatest share of electrical power after coal.[5] Like coal-fired power plants, natural gas-fired plants may be affected by perceived or actual actions to limit GHG emissions; however, they produce markedly lower GHG emissions per unit of electrical output than coal-fired plants. Natural gas-fired power plants are feasible and provide commercially available options for providing electrical generating capacity.

Combined-cycle power plants differ significantly from coal-fired and existing nuclear power plants. They derive the majority of their electrical output from a gas-turbine cycle, and then generate additional power—without burning any additional fuel—through a second, steam-turbine cycle. The first, gas turbine stage (similar to a large jet engine) burns natural gas that turns a driveshaft that powers an electric generator. The exhaust gas from the gas turbine is still hot enough, however, to boil water into steam. Ducts carry the hot exhaust to a heat recovery steam generator, which produces steam to drive a steam turbine and produce additional electrical power. The combined-cycle approach is significantly more efficient than any one cycle on its own; thermal efficiency can exceed 60%. Since the natural gas-fired alternative derives much of its power from a gas turbine cycle, and because it wastes less heat than a coal-fired alternative, it typically requires significantly less cooling.

A.4.4   Nuclear

This section considers both nuclear fission and fusion technologies.

A.4.4.1   Fission

Nuclear energy is generated from a process known as nuclear fission, which splits uranium-235 atoms, releasing large amounts of energy inside a nuclear reactor. A chain reaction results. The heat released from the fission reaction is used to heat water to create steam, which spins a turbine generator, producing electricity.

The economics of nuclear power is a complicated issue, because of high capital costs for building but relatively low fuel costs. Compared with other power generation methods, the analysis is strongly dependent on assumptions about construction timescales and capital financing for nuclear plants. Depending on the source used for claims about energy return on energy investment (EROI), advocates argue that it takes 4–5 months of energy production from the nuclear plant to fully pay back the initial energy investment.[6] In contrast, opponents claim it depends on the grades of fuel ores used, so a full payback can vary from 10 to 18 years, and that the advocates' claim was based on the assumption of high grade ores.[7]

One of the most vexing issues is that of long-term radioactive waste storage. A number of countries have considered using underground repositories.

Depending on the type of fission fuel used, estimates for existing supply of uranium-235 may be limited to several decades at current usage rates. By one estimate, at the present rate of use, there are about 70 years of uranium-235 supply left, assuming economically recoverable uranium at a price of $130/kg.[8] Others argue that the cost of fuel is a minor cost factor, and more expensive, more difficult to extract sources of uranium could be mined in the future, such as granite and seawater;[9] moreover, increasing the price of uranium would have little effect on the total cost of nuclear power; a doubling in the cost of natural uranium would increase the total cost of nuclear power by 5%. In contrast, if the price of natural gas were doubled, the cost of gas-fired electricity would increase by about 60%.[10]

On the other hand, opponents argue that the correlation between price and production is not linear. As uranium ore concentration decreases, the difficulty of mining rises rapidly; the assertion that higher price will yield more uranium has often been inflated; for example, opponents counter that a rough estimate predicts that the extraction of uranium from granite will consume at least 70 times more energy than what it will produce in a reactor. To date, eleven countries have already depleted their uranium resources, and only Canada has mines left which produce better than 1% concentration ore. As a source, seawater seems to be equally dubious. As a consequence, an eventual doubling in the price of uranium may yield only a marginal increase in the volumes that are being produced.

Thorium might provide a suitable fission fuel alternative. Thorium is three times more abundant in Earth's crust than uranium, and

Appendix A

thorium can be bred into Uranium-233, reprocessed, and then used as fuel. But thorium also comes with its own set of power reactor disadvantages.

Breeder reactors can convert the more abundant uranium-238 (99.3% of all natural uranium) into plutonium which can be used as a fuel in nuclear reactors. Fast breeder reactors have been built by several countries. However, these fast breeder reactors have all had difficulties and were not economically competitive, and most have been decommissioned. Perhaps the most vexing problem is that the plutonium can be diverted and used for constructing nuclear weapons.

The possibility of a nuclear meltdown or accident, such as Three Mile Island, Chernobyl, and what was witnessed in Japan's 2011 nuclear disaster, is an ever present problem. Many nuclear experts believe that pebble bed reactors, in which each nuclear fuel pellet is coated with a ceramic coating, are inherently safe and provide the best solution for nuclear power. China has plans to build pebble-bed reactors configured to produce hydrogen for use in powering automobiles.

According to a 2007 study, "World Nuclear Industry Status Report," nuclear energy is presently in a general decline. Table A.2 compares the advantages and disadvantages of this energy system.

A.4.4.2 Fusion—ultimate energy source?

Fusion is the process that powers the Sun and stars. It is capable of producing enormous amounts of energy with almost no radiation or waste byproducts. Moreover, its fuel source can be extracted from the sea and is virtually inexhaustible. Unfortunately, there is a popular statement that fusion has been a mere 20 years away for the last 60 years.

But this is an overstatement. Progress in magnetic confinement of plasmas has been impressive, and quantitative performances, achieved in experiments, have from 1975 outperformed the well known Moore's law of digital computer technology. Seven of the largest nations in the world (United States, China, Europe, Japan, Russia, India, and South Korea) are collaborating on an international project to demonstrate that it is possible to sustain a fusion reaction. The large facility, called ITER, is designed to produce half a gigawatt of fusion power sometime after 2025. This experimental reactor should demonstrate the scientific feasibility of fusion as an energy source; it should validate and optimize the parameters and develop the technologies for the next step, an electricity-generating demonstration reactor for evaluating the economics of fusion. To actualize this will take around 40 years for its design, construction, and to obtain sufficient operating time to capitalize on the results. This could be the ultimate solution to the energy puzzle, but as you can see, it is a long way off.

Table A.2 Advantages and Disadvantages of Fission Nuclear Energy

Advantages	Disadvantages
• The cost of making nuclear power is approximately the same as making coal power, which is considered relatively inexpensive. • Nuclear power does not produce any primary air pollutants or release carbon dioxide and other pollutants. Some carbon dioxide is produced through standard construction practices. Therefore, it contributes only a small amount to global warming. • The energy content of a kilogram of uranium, assuming spent fuel is reprocessed, is equal to 3.5 million kilograms of coal. • According to one study, fast breeder reactors are capable of provide sustainable energy as they have the potential to power society for billions of years. • The life cycle emissions of nuclear power vary; some studies suggest that the emissions are about 4% of those produced by coal power. Depending on the assumptions used, hydro, wind, and geothermal are sometimes ranked lower, while wind and hydro are sometimes ranked higher (by life cycle emissions). • Mining uranium is much safer for nuclear power compared with coal. Coal mining is the second most dangerous occupation in the United States. Nuclear energy is much safer per capita than coal derived energy.	• Nuclear waste is very highly radioactive and very toxic. However, breeder reactors have the potential to "burn" this waste as fuel, transforming the waste into a much more stable and less dangerous waste form. • A major nuclear reactor accident could be catastrophic (i.e., Chernobyl or Fukushima). • Many geologists claim that uranium ore is a limited resource and estimate that current supplies will fail to meet demand by perhaps 2026. This claim is disputed by proponents; breeder reactors would extract about 100 times more energy from the same amount of uranium, greatly extending the supply; however, countries showing an interest in breeder technology are on the decline. • Since nuclear power plants are typically quite large and are, fundamentally, thermal engines, waste heat disposal becomes more difficult. Thus, at peak demand, a power reactor may need to operate at a reduced power level. • There can be a nexus between nuclear power and nuclear weapons proliferation.

A.4.5 *Renewable sources*

Renewable energy, also commonly referred to as *alternative energy*, is an alternative to fossil fuels and nuclear power. Jacobson et. al. advanced a controversial plan to power 100% of the world's energy with solar, wind, and hydroelectric by the year 2030.[11]

A.4.5.1 Wind

Wind energy is harnesses the power of the wind to propel the blades of wind turbines. Wind towers are usually built together on wind farms. Table A.3 compares the advantages and disadvantages of this energy system.

A.4.5.2 Hydroelectric

Hydro energy uses the gravitational descent of a river flow to turn turbines, which drive an electric generator. Hydroelectric dams can result in unexpected results. For instance, one study showed that a hydroelectric dam in the Amazon has 3.6 times more greenhouse emissions per kilowatt-hour (kWh) of electricity generated than from oil, due to the large scale emission of methane from decaying organic material, though this is most significant when river valleys are initially flooded, and is of much less consequence for more boreal dams.[12] In particular, this effect applies to dams created by simply flooding a large area, without first clearing the vegetation. However, underwater turbines that do not require a dam are being investigated. Table A.4 compares the advantages and disadvantages of this energy system.

Table A.3 Advantages and Disadvantages of Wind Energy

Advantages	Disadvantages
• Wind power does not contribute to greenhouse gases.	• Availability of wind is variable and unpredictable. When the wind ceases, no electricity is generated. Thus, wind power is unsuitable for base-load power generation.
• Wind energy produces virtually no air or water pollution. Hence, there are no waste by-products, such as carbon dioxide.	• Wind farms are often opposed by communities that consider them an eyesore or obstruction.
• Farming, grazing, and other activities can still take place on land occupied by wind turbines. Multiple land use can also lower siting costs.	• Depending on the location and type of turbines used, they may pose a danger to the birds (this is an issue of greater concern to older/smaller turbines).
• As a renewable energy source, we will never run out of it.	• Tall wind turbines impact Doppler weather radar towers and affect weather forecasting in a negative way.
• Utilizing wind power in a grid-tie configuration allows for backup power in the event of an outage.	• Wind farms can interfere with radar, creating a hole in radar coverage affecting aviation and national security.
• It can benefit people living in remote areas, where it may be difficult to transport electricity through wires from remote power.	

Table A.4 Advantages and Disadvantages of Hydroelectric Energy

Advantages	Disadvantages
• Hydroelectric power plants can promptly respond to changes in electrical demand, because water can be accumulated above the dam and released to coincide with peak demand. • As long as sufficient water is available, electricity can be generated constantly. • A man-made lake can have additional benefits such as doubling as a reservoir for irrigation, wetlands, and leisure activities. • Hydropower is a renewable energy source. • Hydroelectric energy produces no primary waste or pollution. • A significant potential of hydroelectric capacity is still undeveloped, particularly in some developing nations.	• Hydroelectricity can only be developed in areas where there is a sufficient water supply. • Construction of a dam can result in detrimental environmental impacts. For instance, the amount and quality of water downstream can adversely affect aquatic and terrestrial biota. As a river valley is being flooded, local species can be destroyed, while people living nearby must relocate their homes. • Flooding can submerge large forests. The resulting anaerobic decomposition of the plant material can release methane, a potent greenhouse gas. • Hydroelectric plants often require construction of long transmission lines that have their own adverse effects. • Dams create large lakes that may have adverse effects on the tectonic system causing intense earthquakes. • Dams can fail, leading to catastrophic flooding.

A.4.5.3 Solar

Solar technologies use the sun's energy to produce electricity. Solar power can use:

1. Solar cells to convert sunlight into electricity
2. Sunlight striking solar thermal panels to convert sunlight to heat water or air
3. Sunlight hitting a parabolic mirror to heat water (producing steam)
4. Sunlight entering windows for passive solar heating of a building.

Table A.5 compares the advantages and disadvantages of this energy system.

Electricity demand in the continental United States is approximately 3.7×10^{12} kWh per year. Assuming 20% efficiency, an area of approximately 3,500 square miles would have to be covered with solar panels to replace all current electricity production in the United States. Substantial international investment capital is currently flowing to China to support development of solar energy technology.

Table A.5 Advantages and Disadvantages of Solar Energy

Advantages	Disadvantages
• Solar energy is a renewable resource. • Solar power generation releases little or no air or water pollution. • Solar energy consumes little or no fuel costs. • In sunny areas, solar power can be used in remote locations; isolated places can receive electricity, when there is no way to connect power lines from a plant. • Solar power can be used very efficiently for heating. • Solar energy tends to be abundant in regions that have the largest number of people living off the grid (i.e., developing regions such as Africa and Latin America). • Distributed photovoltaic systems can eliminate expensive long-distance electric power transmission losses. • Photovoltaics are much more efficient in their conversion of solar energy to usable energy than biofuel from plant materials. • Passive solar building designs are demonstrating significant energy bill reduction. • Photovoltaic equipment cost has been steadily falling, and the production capacity is rapidly rising.	• Solar heat and electricity are not available at night and may be unavailable because of weather conditions, therefore requiring expensive storage systems • Solar electricity is currently more expensive than traditional forms of electricity. • The energy payback time (time necessary for producing the same amount of energy as needed for building the power device) for photovoltaic cells is about 1–5 years, depending primarily on location. • The chemicals used to manufacture cells can be extremely hazardous. • Solar cells produce DC which must be converted to AC; this results in an energy loss of 4–12%.

A.4.5.4 Biomass

Biomass energy production uses garbage or other renewable resources such as vegetation to generate electricity. As garbage decomposes, it produces methane, which is captured in pipes and later is burned to produce electricity.

Vegetation and wood can be burned directly to generate energy, as in the case of fossil fuels, or processed to form alcohols. Straight vegetable oil works in diesel engines if it is heated first. Vegetable oil can also be treated to make biodiesel, which burns like normal diesel. Table A.6 compares the advantages and disadvantages of this energy system.

In addition to wood and municipal solid waste fuels (described below), which are proven sources of energy, there are other concepts for biomass-fired electric generators, including direct burning of energy crops, conversion to liquid biofuels, and biomass gasification; however,

Table A.6 Advantages and Disadvantages of Biomass Energy

Advantages	Disadvantages
• Biomass can be burned to produce energy. • Biomass is abundant and renewable. • When methods of biomass production other than direct combustion of plant mass are used, such as fermentation of alcohol, it results in few environmental effects. Alcohols tend to burn relatively cleanly and are feasible replacements for fossil fuels. • Because CO_2 is first taken out of the atmosphere to make the vegetable oil and then returned as it is burned in an engine, there is no net increase in CO_2. However, there are still emissions as a result of fossil fuel used in growing and producing biofuel. • Vegetable oil can be easily transformed into biodiesel and used as a replacement for diesel. • Has the potential to produce far more vegetable oil per acre than current plants. • The infrastructure for biodiesel is significant and growing around the world.	• Direct combustion of any carbon-based fuel leads to some air emissions similar to that from fossil fuels. • Even under the most-optimistic energy return on investment studies, using 100% solar energy to grow corn and produce ethanol (fueling machinery with ethanol, distilling with heat from burning crop residues, using no fossil fuels at all), the consumption of ethanol to replace only the current U.S. petroleum use would require three quarters of all the cultivated land on the face of the Earth. • Some studies indicate that when biomass crops are raised through intensive farming, the production of ethanol fuel results in a net loss of energy after one accounts for the fuel costs of petroleum and natural-gas fertilizer production transportation, and distillation process. • Current production methods require large amounts of land. With current technology, it is not feasible for biofuels to replace the total demand for petroleum. • Competes with land use for food production and water use. This decreases food supply, and raises the price of food worldwide.

none of these other technologies has progressed to the point of being competitive on a large scale or of being reliable enough to replace the baseload of a conventional electrical generation plant.

A.4.5.5 Wood waste

The Bioenergy Feedstock Development Program at Oak Ridge National Laboratory estimated that each air-dried pound of wood residue produces approximately 6,400 Btu of heat. Assuming a 33% conversion efficiency, biomass might be capable of generating 30.3 tWt hours of electricity. This study goes on to note that these estimates of biomass capacity contain substantial uncertainty, and that potential availability does not mean biomass

Appendix A 149

would actually be available at the prices indicated or that resources would be usably free of contamination. Some of these plant wastes already have alternative uses and would likely be more costly to deliver because of competition. Others, such as forest residues, may prove unsafe and unsustainable to harvest on a regular basis.

A.4.5.6 Municipal solid waste
Municipal solid waste combustors use three types of technologies—mass burn, modular, and refuse-derived fuel. Mass burning is currently the method used most frequently in the United States and involves no (or little) sorting, shredding, or separation. Consequently, toxic or hazardous components present in the waste stream are combusted, and toxic constituents are exhausted to the air or become part of the resulting solid wastes. Currently, approximately 89 waste-to-energy plants operate in the United States. These plants generate approximately 2,700 MWe, or an average of 30 MWe per plant.

Waste-fired plants have the same or greater operational impacts than coal-fired technologies (including impacts on the aquatic environment, air, and waste disposal). The initial capital costs for municipal solid-waste plants are greater than for comparable steam-turbine technology at coal-fired facilities or at wood-waste facilities because of the need for specialized waste separation and handling equipment.[13]

The decision to burn municipal waste to generate energy is usually driven by the need for an alternative to landfills rather than energy considerations. The use of landfills as a waste disposal option is likely to increase in the near term as energy prices increase; however, it is possible that municipal waste combustion facilities may become attractive again.

A.4.5.7 Geothermal
Geothermal power harnesses heat energy present within the Earth. Typically, two wells are drilled, one to inject water into the subsurface. Once injected, the hot rocks heat the water to produce steam. The steam shoots back through the second hole and is used to drive turbines that power electric generators. There are also natural sources of geothermal power (hot springs, volcanoes, geysers, and steam vents). However, geothermal electric generation is limited by the geographical availability of geothermal resources.[14] Nevertheless, some studies have concluded that geothermal is one of the largest potential sources of renewable energy. Table A.7 compares the advantages and disadvantages of this energy system.

A.4.5.8 Wave and ocean energy
Ocean waves, currents, and tides are often predictable and reliable. Ocean currents flow consistently, while tides can be predicted months and years in advance because of their well-known behavior.

Table A.7 Advantages and Disadvantages of Geothermal Energy

Advantages	Disadvantages
• It is currently economically feasible in high grade areas. • The deployment costs are relatively low. • Geothermal energy provides a base load form of power. The power plants have a high capacity factor (i.e., they run continuously day and night with an uptime typically exceeding 95%). • Geothermal energy produces little air or water pollution. • Because geothermal power stations consume no fuel, there is little environmental impact associated with emissions. • Geothermal is currently feasible in areas where the Earth's crust is thicker; using enhanced geothermal technology, it is possible to drill deeper and inject water to generate geothermal power. • Once a power station is constructed, there is no cost for fuel, only for operations, maintenance, and return on the capital investment. • In many circumstances it is considered to be a renewable energy resource. • It provides a near constant and reliable source of energy.	• Geothermal energy extracts small amounts of pollutants, such as sulfur, that need to be removed prior to feeding the turbine and reinjecting the water back into the injection well. • Some geothermal plants have created geological instability, including small earthquakes strong enough to damage some buildings. • Geothermal energy requires locations that have suitable subsurface temperatures within 5 km of the surface.

Tidal energy can be extracted from tides by (1) locating a water turbine in a tidal current, or (2) building impoundment pond dams that admit or release water through a turbine. The turbine can turn an electrical generator or a gas compressor that can then store energy until needed. Tidal power tends to be a source of clean, free, and sustainable energy. However, most of these technologies are in relatively early stages of development. Table A.8 compares the advantages and disadvantages of this energy system.

A.4.5.9 Hydrogen

Hydrogen can be manufactured at roughly 77% thermal efficiency through *steam reforming* of natural gas. Hydrogen can also be used as (1) a fuel cell, which converts the chemicals hydrogen and oxygen, into water and, in the process, produces electricity, or (2) hydrogen can be burned (less efficiently than in a fuel cell) in an internal combustion engine. There are three principal ways of creating hydrogen:

Appendix A

Table A.8 Advantages and Disadvantages of Tidal Energy

Advantages	Disadvantages
• Tidal energy is free once the plant is built, because tidal power harnesses the natural power of tides and does not consume fuel. • The maintenance costs of running a tidal plant are relatively low. • Tides are very reliable; the tide goes in and out twice a day at predicted times. This makes tidal energy easy to maintain, such that positive and negative energy spikes can be managed. • Tidal power is renewable because it relies on the gravitational pull of the moon and sun, which pull the sea backwards and forwards, generating tides.	• Tidal energy is not currently economical on a wide-scale basis, because the initial costs of building a dam are large. Moreover, it only provides power for approximately 10 hours per day, when the tide is moving in or out of the basin. • Boats may not be able to cross the barrage, and commercial ships used for transport or fishery may need to find alternative routes. • Construction of a barrage can affect the aquatic ecosystems. • Maximum energy production is limited to about 2.5 tWs (total amount of tidal dissipation or the friction measured by the slowing of the lunar orbit).

1. By breaking down hydrocarbons such as methane using steam reforming. If petroleum is used to provide the energy for this process, fossil fuels are consumed, forming pollution and nullifying most of the value of using a fuel cell.
2. Through the electrolysis of water, which involves splitting water into oxygen and hydrogen using electrolysis. However, current processes may require more energy to produce hydrogen through electrolysis than one can obtain from hydrogen. It has been calculated that it takes 1.4 joules of electricity to produce 1 joule of hydrogen.
3. By reacting water with a metal such as sodium or potassium, and producing toxic waste byproducts.

Table A.9 compares the advantages and disadvantages of this energy system.

A.4.5.10 *Battery storage systems*

Batteries can store energy in a chemical form. Batteries can be used to store energy in battery-powered electrical vehicles. These electric vehicles can be charged from the grid. However, the electrical energy to charge the battery must come from a source such as coal or gas-fired power plants, wind, solar, geothermal, hydroelectric or nuclear plants. Table A.10 compares the advantages and disadvantages of this energy system.

Table A.9 Advantages and Disadvantages of Hydrogen Energy

Advantages	Disadvantages
• Hydrogen is virtually nonpolluting, yielding pure water vapor as exhaust when combusted in air. • Hydrogen can be produced domestically from the decomposition of water. • Hydrogen is the lightest chemical element and has the best energy-to-weight ratio of any fuel.	• With minor exceptions, hydrogen does not exist in its pure form in the environment, because it reacts so strongly with oxygen and other elements. • It is impossible to obtain hydrogen gas without expending energy in the process. • There is currently modest infrastructure for distribution of hydrogen. It would require massive injections of capital and decades to convert our current technological society into a hydrogen-based economy. • Hydrogen is difficult to handle, store, and transport. • Hydrogen is extremely explosive.

Table A.10 Advantages and Disadvantages of Battery Storage Systems

Advantages	Disadvantages
• Electric motors are about 90% efficient compared to about 20% efficiency for an internal combustion engine. • Current lead acid battery technology offers a 50+ mile range on one charge. • The use of battery electric vehicles may reduce the dependency on fossil fuels. • Battery-powered electric vehicles have fewer moving parts, thus improving the reliability. • The cost of purchasing a battery-powered vehicle is currently substantially higher than a conventional vehicle. • The cost of operating a battery powered electric vehicle is approximately 2 to 4 cents per mile, or about a sixth the price of a gasoline powered vehicle. • Battery electric vehicles are relatively quiet. • They produce no direct hazardous emissions of Co_2. However, plants that generate electric power used for charging batteries may produce such emissions.	• Key materials required in battery production, such as lithium, are becoming very scarce • Current battery technology is expensive. • Grid infrastructure would need to be improved significantly to accommodate mass-adoption of grid-charged electric vehicles. • Battery-powered electric vehicles have a relative short range and require long recharge times. • Some batteries perform poorly in cold weather while other types perform poorly in hot weather. • Some batteries are highly toxic.

A.4.5.11 Fuel cells

Fuel cells oxidize fuels without combustion and its associated environmental side effects. Power is produced electrochemically by passing a hydrogen-rich fuel over an anode and air (or oxygen) over a cathode and separating the two by an electrolyte. The only byproducts (depending on fuel characteristics) are heat, water, and CO_2. Hydrogen fuel can come from a variety of hydrocarbon resources by subjecting them to steam under pressure. Natural gas is typically used as the source of hydrogen.

At the present time, fuel cells are generally not economically or technologically competitive with other alternatives for electricity generation. Fuel cells may cost $5,374 per installed kilowatts (kW) (total overnight costs), or 3.5 times the construction cost of new coal-fired capacity and 7.5 times the cost of new, advanced gas-fired, combined-cycle capacity.[15] In addition, fuel cell units are likely to be small in size.

Endnotes

1. Gorman, Siobhan (2009-04-08). Gorman, Siobhan (2009-04-08). "Electricity Grid in U.S. Penetrated By Spies," *The Wall Street Journal*.
2. Kannberg, L. D., Kintner-Meyer, M. C., Chassin, D. P., Pratt, R. G., DeSteese, J. G., Schienbein, L. A., Hauser, S. G., Warwick, W. M. (2003–11) (pdf). *GridWise: The Benefits of a Transformed Energy System*. Pacific Northwest National Laboratory under contract with the United States Department of Energy. p. 25.
3. National Energy Technology Laboratory (2007-07-27) (pdf). *A Vision for the Modern Grid*. U.S. Department of Energy. p. 5.
4. Energy Information Administration (EIA). 2009. "Summary Statistics for the United States," Table ES1 from *Electric Power Annual with Data for 2007*. Available URL:http://www.eia.doe.gov/cneaf/electricity/epa/epates.html (accessed June 2009).
5. Energy Information Administration (EIA). 2009. "Summary Statistics for the United States," Table ES1 from *Electric Power Annual with data for 2007*. Available URL:http://www.eia.doe.gov/cneaf/electricity/epa/epates.html (accessed June 2009).
6. "Energy Analysis of Power Systems". World Nuclear Association. March 2006.
7. "Coming Clean; How Clean is Nuclear Energy?" October 2000. "World Information Service on Energy" 10–18 years for payback on nuclear energy, Jan Willem Storm van Leeuwen; Philip Smith (July 39, 2005). "Nuclear Energy: the Energy Balance."
8. "Supply of Uranium." World Nuclear Association. March 2007.
9. "The Economics of Nuclear Power." World Nuclear Association. June 2007.
10. "The Economics of Nuclear Power." World Nuclear Association. June 2007.
11. Jacobson, M.Z. and Delucchi, M.A. (November 2009) "A Plan to Power 100% of the Planet with Renewables" (originally published as "A Path to Sustainable Energy by 2030"). *Scientific American* 301(5):58–65.

12 Graham-Rowe, Duncan (2005-02-24). "Hydroelectric power's dirty secret revealed." *New Scientist*.
13 U.S. Nuclear Regulatory Commission (NRC). 1996. *Generic Environmental Impact Statement for License Renewal of Nuclear Plants*, NUREG-1437, Volumes 1 and 2, Washington, D.C.
14 U.S. Nuclear Regulatory Commission (NRC). 1996. *Generic Environmental Impact Statement for License Renewal of Nuclear Plants*, NUREG-1437, Volumes 1 and 2, Washington, D.C.
15 Energy Information Administration (EIA). 2009. "Summary Statistics for the United States." Table ES1 from *Electric Power Annual with Data for 2007*. Available URL:http://www.eia.doe.gov/cneaf/electricity/epa/epates.html (accessed June 2009).
16 "Environmental Impacts of Coal Power: Air Pollution." Union of Concerned Scientists.

Appendix B: Perspectives on energy efficiency and conservation

Though often used interchangeably, energy conservation and energy efficiency are different concepts. *Energy efficiency* typically means deriving a similar level of services by using less energy, while *energy conservation* simply indicates a reduction in energy consumption; it is possible to increase energy conservation without reducing total energy consumption.

Modern society has benefited from exploiting energy sources, particularly since energy became much cheaper and more abundant during the Industrial Revolution. Our society is now deeply dependent on cheap and abundant sources of energy. Nevertheless, it is important to note that "energy consumption" does not directly improve the human condition. Instead, what matters are the services that are derived from energy generation. As Lovins has pointed out:

> Customers don't want lumps of coal, raw kilowatt-hours, or barrels of sticky black goo. Rather, they want the services that energy provides: hot showers and cold beer, mobility and comfort, spinning shafts and energized microchips, baked bread and smelted aluminum. And they want these "end uses" provided in ways that are secure, reliable, safe, healthful, fair, affordable, durable, flexible, and innovation friendly.

But this observation extends to more than just the energy sector. Our society also consumes large amounts of material resources (water, minerals, metals, and chemicals). One of the principal concerns associated with material resources involves the environmental impacts and energy used in extracting, transporting, and forging them into useful products.

Energy efficiency is not "pie in the sky." For instance, due to federal and state efficiency regulations, refrigerators today consume 75% less energy that they did back in 1972. According to Vivian Loftness at Carnegie Mellon University, residential and commercial energy use consumes 72% of all electricity and 13% of fossil-fueled energy. This implies that buildings offer enormous potential for energy savings. For instance, natural daylight can replace 30–60% of electrical lighting, and natural ventilation can reduce electrical air-conditioning consumption by 20–40%, while natural shading can account for an additional 10% in energy consumption, and passive heating could eliminate 20–40% in heating costs. Conservation essentially substitutes for an entirely new source of energy. Loftness goes on to state that stringent energy regulations or perhaps a rigorously enforced energy management system (EnMS) could save 50–75% on the cost of running equipment and appliances.[1] With such potential savings, it is amazing that more action has not been focused on energy efficiency until now.

According to Lowell Ungar, director of Alliance to Save Energy, buildings, account for 20% of our energy consumption and carbon dioxide (CO_2) emissions. His data indicate that stronger building codes, such as improved insulation and lighting, could reduce building energy consumption by 6–7% by 2030. Building in dilapidated city centers could save gasoline by reducing urban sprawl and the distance needed to commute to remote urban sites.

Proper labeling on consumer products such as the U.S. EnergyStar program can inform consumers about the energy consumption and cost of operating equipment and appliances. Energy efficiency could be forced through tax incentives or draconian regulations and standards imposed on appliances, equipment, buildings, and vehicles by bureaucrats. However, there is another approach: voluntary but rigorously implemented adoption of EnMSs by corporations, organizations, and government agencies that would allow them the freedom to develop their own energy saving polices and plans tailored to their specific needs.

Of all the available statistics, perhaps no other demonstrates the energy problem we are facing more than the following. Despite advances in efficiency and sustainability, of all the energy created since the Industrial Revolution, more than half has been consumed in the last two decades.[2]

In 2009, world energy consumption decreased for the first time in 30 years (–1.1%) or 130 Mtoe, as a result of the financial and economic crisis (GDP drop by 0.6% in 2009).[3]

B.1 Energy conservation

An improvement in energy production and use is readily achievable. In the industrial sector, increasing energy costs have forced numerous

Appendix B: Perspectives on energy efficiency and conservation

energy-intensive industries to make substantial efficiency improvements since the 1970s. For example, the energy consumed in producing steel has been reduced by nearly 40%; to date, most of this reduction in energy usage has resulted from recycling waste materials and the use of cogeneration equipment.

However, economic theory (Jevon's Paradox) suggests that technological improvements increasing energy efficiency actually tend to increase rather than reduce energy use (See Figure B.1). While somewhat controversial, this paradox involves the proposition that technological progress which increases resource efficiency tends to increase (rather than decrease) the rate of consumption of that resource. Thus, increased technological efficiency may actually lead to long-term increases in the cost of energy due to increased demand. The implications of this paradox are that technological improvements and increased energy efficiency may not be the ultimate solution to the world's energy woes. Figure B.1 illustrates this point. In this example, improved technology (fuel usage) causes the amount of work produced by a given amount of fuel to double (increased efficiency).

This causes the cost of fuel to drop in half. However, because the cost of fuel drops in half, the total fuel usage (demand) more than doubles. Thus, instead of reducing total fuel consumption, more efficient technology has increased demand for fuel, the exact opposite of what is desirable—a paradox.

The solution may actually rest in energy management which enables organizations to take a systematic and rigorously managed approach to achieving continual improvement of energy performance, energy conservation, and energy efficiency.

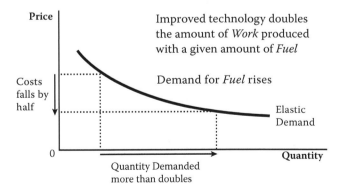

Figure B.1 Jevon's Paradox. Improved technology doubles the amount of *work* produced with a given amount of *fuel*.

B.2 Energy use and efficiency around the world

Energy usage continues to increase the world over. The social, economic, geopolitical, and environmental impacts increase in tandem with this increased usage. Perhaps no resource depicts this problem better than the soaring demand for petroleum. Figure B.2 compares the exponential increase in global population with the exponential increase in global oil production (See Appendix C). As demand for petroleum tightens, we can expect to see socioeconomic, environmental, and geopolitical risks increase accordingly. Clearly, both the increase in population and oil depicted in Figure B.2 are unsustainable.

Figure B.3 shows how various countries compare in terms of the cost of producing a British thermal unit (BTU) of energy per dollar (in year 2000 dollars). Among the nations depicted in Figure B.3, Japan is the least efficient nation depicted, while China is the most efficient. The U.S. ranks somewhere in between these two extremes. Denmark and the U.S. are the first countries to develop national energy management standards (year 2000). Despite its large investment in new technologies, the U.S. still underperforms many other nations. This figure suggests that with the implementation of an effective EnMS, there is substantial room for making significant improvements.

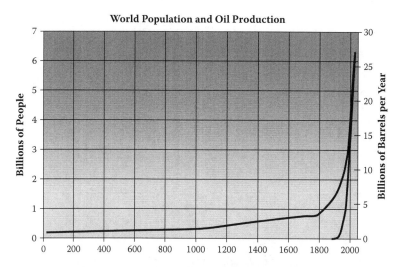

Figure B.2 The plot shows the relationship between the increase in global population and oil production per year. Global population began to soar around 1800, while global oil production began to soar around the mid-1800s. Such exponential growth is widely considered to be unsustainable.

Appendix B: Perspectives on energy efficiency and conservation 159

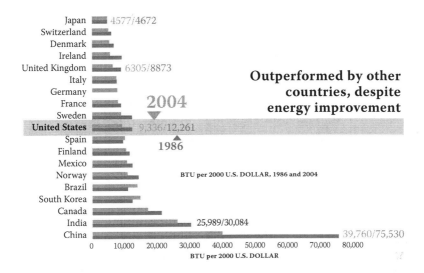

Figure B.3 Energy use and efficiency around the world. (Courtesy of Council on Competitiveness. 2007. *Competitiveness Index: Where America Stands,* Figure 4.32, p. 103).

Industry alone cannot control prices or politics. However, industries have a new and powerful tool for managing their use of energy. The U.S. Department of Energy's (DOE) Industrial Technologies Program, Industrial Sector National Initiative, has set a goal of achieving a 25% reduction in industrial energy intensity within a next decade. Yet, until now, no internationally accepted EnMS has existed for providing direction on reaching such an aggressive target.

According to James D. McCalley, an electrical engineer at Iowa State University, a stunning 57% of all energy is wasted. Investing in ways to conserve energy would be equivalent to developing a major and entirely new source of energy.[4]

No energy source epitomizes the problems that we face more than do fossil fuels. The very essence of modern technology is dependent on cheap, abundant, and reliable fossil fuels. Without a sufficient supply of fossil fuels, the lion's share of our present technology would draw to a halt. Figure B.4 shows the current breakdown and daunting problems associated with substituting renewable and alternative energy sources for fossil fuels. This figure clearly shows why adoption of an EnMS by organizations is needed as a means of reducing energy usage and identifying alternative energy measures, particularly those that reduce dependency on fossil fuels.

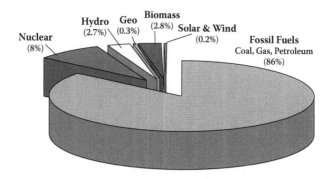

Figure B.4 Comparison of renewable and alternative energy sources with that of fossil fuels.

B.3 Energy conservation/energy efficiency issues

Both energy efficiency and energy conservation fall into a larger category known as demand-side management (DSM). DSM measures—unlike the energy supply alternatives discussed in previous sections—address energy end uses. DSM can include measures that shift energy consumption to different times of the day to reduce peak loads, measures that can interrupt certain large customers during periods of high demand or measures that interrupt certain appliances during high demand periods, and measures like replacing older, less efficient appliances, lighting, or control systems. It also includes measures that utilities use to boost sales, such as encouraging customers to switch from gas to electricity for water heating.

New technology makes better use of energy through improved efficiency, such as more efficient fluorescent lamps and insulation. Energy efficiency shows great promise and has been increasing by about 2% a year.

Energy efficiency and conservation examples abound: Heat exchangers make it possible to recover heat from warm water and air. Production of hydrocarbon fuel from pyrolysis can also be used, allowing recovery of some of the energy in hydrocarbon waste. Power plants often can be made more efficient with minor technological improvements. New power plants use efficient technologies like cogeneration. New building designs use techniques like passive solar heating and lighting. Light-emitting diodes can replace inefficient light bulbs.

Mass transportation can greatly increase energy efficiency compared with the conventional automobile. Learning more efficient ways to operate the conventional combustion engine automobile has continually

Appendix B: Perspectives on energy efficiency and conservation   161

Table B.1 Controversial Energy Conservation Issues

- Some economists argue that bright lighting stimulates purchasing. But health studies show that this can lead to adverse health effects (stress, headache, blood pressure, fatigue). Natural daylighting has been shown to increase worker productivity, while reducing energy consumption.
- Economic theory dictates that technological improvements increase energy efficiency, rather than actually reduce energy usage (Jevons Paradox). This paradox results from two causes: (1) Increased energy efficiency makes the use of energy relatively cheaper, thus encouraging increased use, and (2) increased energy efficiency leads to more economic growth, which increases energy use across the entire economy. To reduce energy consumption, efficiency gains may be paired with government intervention aimed at reducing demand or other mechanisms such as an EnMS that establishes an energy policy and provides a mechanism for ensuring that it is implemented.
- Use of telecommuting by major corporations is a significant opportunity to conserve energy, yet many employers refuse to consider new approaches for which they have little or no experience.
- Electric motors consume more than 60% of all electrical energy generated and are responsible for the efficiency loss of 10 to 20% of all electricity converted into mechanical energy.
- Modern real-time energy metering (Energy Detective, Enigin Plc's Eniscope, Ecowizard, or solutions like EDSA'a Paladin Live) can allow consumers and workers to save energy by monitoring the energy expenditure of their actions.
- Consumers tend to be poorly educated about the potential savings of energy efficient products. Some organizations and governments are attempting to reduce this complexity with "ecolabels" by making differences in energy efficiency easy to comprehend.

improved savings, such as recovering energy spent during braking or turning off the motor when idling in traffic. Hybrid vehicles show great promise. Electric vehicles such as Maglev are much more efficient than combustion based systems.

Both advocates and critics debate a number of energy conservation issues such as those depicted in Table B.1.

Endnotes

1. Loftness V., *Discovery Magazine*, p. 48, June 2010.
2. "Historical Review of Energy Use." ND. http://www.wou.edu/las/physci/GS361/electricity%20generation/HistoricalPerspectives.htm. Retrieved 2010-11-05.
3. "Global Energy Review in 2009," *Enerdata*.
4. McCalley J.D., *Discovery Magazine*, p. 48, June 2010.

Appendix C: Peak oil: The looming oil crisis*

The United States now imports over 50% of the oil it consumes, and this trend is increasing. Transportation accounts for two-thirds of the oil consumed in the United States.

With less than 5% of the world's population, about 35% of the world's vehicles travel down American roads, racking up as many miles each year as the rest of the world combined. Many other Western nations do not fare much better.

This scenario is clearly unsustainable. So why doesn't the West wean itself of dependency on imported oil? The principal reason is the current "abundance" of cheap underpriced oil. Moreover, the cost of pollution is largely not factored into the price of fuel, making it appear to be a bargain.

But there are other factors accounting for this energy dilemma as well. Case in point: By the 1920s, 20 million Americans used trolleys and streetcars. But this was not to last. In the 1940s, General Motors, Standard Oil of California, Firestone, and Phillips Petroleum formed a holding company known as National City Lines. This company purchased privately owned streetcar systems in some 100 major American cities. These older systems were slowly dismantled and replaced by buses and cars that were promoted by the holding company. Then the bus systems were allowed to fail, which created an increased demand for automobiles. This, of course, increases the sale of automobiles, oil, and rubber. These companies were taken to court. The companies were found guilty of conspiracy to eliminate 90% of the American light rail system. The corporate executive officers were each fined $1 and the companies were each fined $5000. In the end, General Motors had profited $25 million on this racket, not including how much it made on increased bus and car sales.[1]

---

* This chapter is based on Chapter 11 of the author's textbook, Global Environmental Policy: Concepts, Principles, and Practice (Eccleston & March) (2010).

The American economy and much of the West is overly dependent on motor vehicles. It has been estimated that one-sixth of every dollar and one-sixth of every nonfarm job is related to motor vehicles. Motor vehicles account for one-fourth of the American national trade deficit.[2] But as you will see in this chapter, their true impacts could be much greater than simply pollution or contributing to the national trade deficit.

C.1 King Hubbert's prediction

When the American Petroleum Institute met for a conference in March 1956, most of the participants expected to hear upbeat news. After all, back in the 1950s, the U.S. petroleum industry was humming and vibrant. The petroleum companies were producing more U.S.-generated crude oil than ever before—in excess of 7 million barrels per day.[3] The sky was the limit. Or so it appeared in March of 1956. Then the brilliant M. King Hubbert stepped up to deliver his paper. His audience did not greet his predictions with enthusiasm.

Dr. Hubbert, then age 52, was a brilliant geologist with Shell Oil Company. Top officials at his company had received advance word of Hubbert's speech and were worried. It is said that the president of the company personally pleaded with him to downplay some of the paper's more controversial claims.[4]

But Hubbert refused and went ahead with his presentation, entitled "Nuclear Energy and the Fossil Fuels."[5] His theme was grim, and it disturbed many of his distinguished guests: He boldly predicted that U.S. oil production would peak somewhere around 1970, and then begin declining.

C.1.1 Peak oil theory

Hubbert argued that U.S. petroleum production would follow a statistical bell-shaped curve. He noted that the quantity of oil available for production in any given region must necessarily be finite, and therefore subject to depletion at some point.

Whenever a new oil field is discovered, the petroleum yield from that location tends to increase rapidly for a period of years, as drilling infrastructure is put in place and extraction activities are ramped up. Once approximately half of the oil field's reserves are pumped out, however, the oil source reaches its peak rate of production. Then decline sets in, with the rate of production decrease ultimately approximating a "mirror image" of the production increase rate seen in the oil field's early years.

The actual date when oil production reaches its peak depends on the number of barrels extracted from the oil field. So plugging different figures into the variables of Hubbert's equations can yield a range of "peak

Appendix C: Peak oil: The looming oil crisis 165

date" estimates. The end result is the same, however: Oil production will eventually crest and then begin a decline—sometimes a rapid descent.

C.1.2 Trajectory of U.S. oil production

In the years following Hubbert's speech, U.S. oil production continued its steady upward progression. The year 1970 came and went. Many commentators dismissed Hubbert. But trends are not always immediately apparent. It can take a few years of data before they become clear. And soon enough, the statistics were in, and they were indisputable: Hubbert had been right—dead on. U.S. oil production had, in fact, peaked in 1970—exactly the year he had predicted, based on his "high oil inventory" estimate.[6]

C.1.3 Repeating the peak?

But Hubbert's prediction went much further. He argued that his theory applied not only to the U.S. petroleum industry, but to global oil production as well. Throughout the 1960s and 1970s, Hubbert continued to refine his theory. He eventually predicted that a peak, also referred to as "Peak Oil," in world oil production, would occur around 2000.[7] The concept of Peak Oil is critical to our modern technological society. Once the peak is reached, the world will be at the highest production level it will ever witness. Oil production will generally decline from that point onward. Adding to this problem is the fact that the decline will occur as world demand is increasing. This will have grave implications for the world economy, agricultural production, travel, and security. For instance a temporary gap of only 5% between supply and demand in 1973 caused the price of gasoline to more than double. What will happen following a future global peak, as the gap begins to increase to 10, 15, or 20%?

The graph in Figure C.1 shows a forecast that Hubbert made in the late 1960s, based on estimated world oil supplies. Assuming the production rates shown in this graph, the world would deplete about 80% of its available oil in a period of under 65 years from the date the graph was first constructed.

C.1.4 Developments in world oil production

After Hubbert made his prediction about global Peak Oil, political events arose that affected his estimates. In particular, the 1973 oil embargo imposed by the Organization of the Petroleum Exporting Countries (OPEC) resulted in an "energy crisis" that changed the dynamics of Hubbert's equation. By reducing oil consumption and encouraging greater energy efficiency, the embargo and its aftermath probably delayed the worldwide Hubbert's Peak.[8]

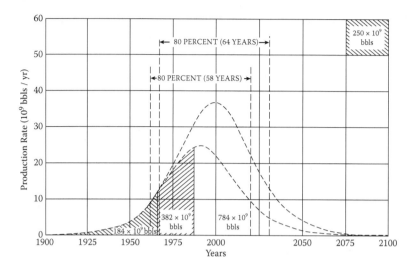

Figure C.1 Bell-shaped curve of global oil production, known as the Hubbert Peak, was published by King Hubbert in 1969.

C.1.5 Heading into the future

While a number of factors may have affected Hubbert's projections, the point of a global peak is still inevitable. Some Peak Oil specialists believe that world production is currently in the process of peaking as this book is being written.

Every oil field has its own unique Hubbert Curve. Once its peak is reached, decline is inevitable. To see what this means in practice, consider oil production in Pennsylvania, where the U.S. petroleum industry began in the mid-19th century. Pennsylvania's production has been in decline for decades. There are approximately 20,000 producing oil wells in this state. These wells each yield an average of only around one-quarter barrel per day. Once a mighty producer of crude, today all the oil wells in Pennsylvania combined produce less than half as much as a single high-producing well in Saudi Arabia.[9]

Thousands of oil fields around the world have already witnessed their peak production. Thousands more are quickly heading toward the same fate as those in Pennsylvania. And even where production is not yet clearly in decline, petroleum may not be as plentiful as it appears, or as we hoped.

C.1.6 Reserve estimates—Are they trustworthy?

Assumptions about the amount of oil available to be pumped are based largely on estimates of oil reserves announced by petroleum producers.

But many observers suspect that producer estimates are inflated.[10] These observers note that between 1980 and 1990, the "proven reserves" of many of the largest oil-producing nations suddenly and mysteriously spiked in a dramatic fashion. For example, Abu Dhabi's declared reserves mysteriously tripled in single year, from approximately 31 billion barrels in 1987 to 92 billion barrels in 1988. That same year, Venezuela's declared reserves more than doubled, from 25 billion to over 56 billion barrels. Many OPEC nations followed in lockstep.

What could account for such sharp and massive increases in reserves? The answer may have little to do with the actual amount of oil in the ground—but a lot to do with OPEC politics. Since 1985, OPEC has tied member countries' production quotas to their oil reserves. Under this policy, members are allowed to pump only the quantity of oil specified in their individual quotas, and these quotas are based on the estimated reserves that each OPEC member declares. So the higher a country's reserve estimates, the more oil it is allowed to pump—and the more revenue it can earn from petroleum sales. This of course creates a strong incentive for OPEC states to overstate their reserves in order to gain larger production quotas. After all, the bigger the production quota, the more income the country can generate.

Kuwait was the first OPEC nation to suddenly increase its reserve estimate, which soared dramatically by over 40% in a single year (1984 to 1985). Within a few years, several other nations (including oil giant Saudi Arabia) responded by dramatically increasing their officially stated reserves. These developments strongly suggest that the estimated reserves of many major oil-producing nations have been purposefully and greatly exaggerated.

Moreover, many OPEC members' reserve estimates are not declining appreciably from one year to the next. This means that OPEC members are in effect claiming to discover "new" fields—year after year—that coincidently match (and replace) the quantities of oil they are pumping out.[11] The overwhelming consensus of the geological community is that oil consumed today was formed millions of years ago through biological and geochemical processes. The Earth's petroleum reserves are therefore essentially finite and declining.

If OPEC producers are in fact overstating their reserves, then the amount of oil yet to be extracted may actually be much smaller than official projections estimate. This means that the "day of reckoning" may be much closer than previously thought. In fact, some Peak Oil critics believe that oil production by many OPEC members is already in the process of peaking. Now here's the problem: Hubbert's equations—indeed, his predictions—are predicated on the volume of oil that remains to be pumped out. If the estimates are exaggerated, the date in which world crude production peaks will be shortened. It is simply a matter of time.

C.2 Growing oil consumption and lower exports by producers

The problem is made worse by growing rates of oil consumption in many countries around the world—including petroleum-producing nations themselves. A recent article in the *New York Times* noted, "The economies of many big oil exporting countries are growing so fast that their need for energy within their borders is crimping how much they can sell abroad, adding new strains to the global oil market."[12] The article went on to add[13]:

> Experts say the sharp growth, if it continues, means several of the world's most important suppliers may need to start importing oil within a decade to power all the new cars, houses, and businesses they are buying and creating with their oil wealth.
> Indonesia has already made this flip. By some projections, the same thing could happen within five years to Mexico, the No. 2 source of foreign oil for the United States, and soon after that to Iran, the world's fourth-largest exporter.

C.2.1 A "solution" to peak oil?

We are unlikely to explore or drill our way out of this pending crisis. As Peak Oil geologist Colin Campbell has stated:

> The whole world has now been seismically searched and picked over. Geological knowledge has improved enormously in the past 30 years, and it is almost inconceivable now that major fields remain to be found.

Many who dismiss Peak Oil predictions—particularly some economists—argue that improved technologies and higher oil prices will solve any oil supply problems that may arise. But many petroleum geologists view the situation differently. They point out that traditional economic incentives may not work when the underlying problem is a natural resource shortage. Higher prices cannot make oil appear if it does not exist. Moreover, feverish oil drilling in areas of mediocre potential and other "last minute" efforts are unlikely to provide a "lifeboat" once production peaks.

C.2.2 U.S. department of energy's peak oil assessment

A 2005 study of the looming Peak Oil problem was commissioned by the U.S. Department of Energy (DOE). The results of this study paint a sobering picture of the problem—and the level of effort needed to address it. The study (Hirsch Report) resulted in a report entitled "Peaking of World Oil Production: Impacts, Mitigation, and Risk Management."[14]

The report's executive summary opened with an ominous sentence: "The peaking of world oil production presents the U.S. and the world with an unprecedented risk management problem."[15] Table C.1 summarizes some of the key conclusions made in the Hirsch Report.

The authors note, "Without mitigation, the peaking of world oil production will almost certainly cause major economic upheaval. However, given enough lead-time, the problems are solvable with existing technologies."[16] The authors go on to say that the "obvious conclusion" from their overall analysis "is that with adequate, timely mitigation, the costs of peaking can be minimized. If mitigation were to be too little, too late, world supply/demand balance will be achieved through massive demand destruction (shortages), which would translate to significant economic hardship."[17]

C.3 Peak oil policy implications

As pointed out earlier, the Hirsch Report notes that as the day of peaking approaches, excess production capacity will dwindle, such that even minor

Table C.1 Key Conclusions Made in the Hirsch Report

- World oil production will peak, although experts differ on exactly when the peak will occur.
- Peak Oil will have a severe impact on the U.S. economy.
- Peak Oil is a "unique challenge," something the world has never before faced. The authors note, "Previous energy transitions (wood to coal and coal to oil) were gradual and evolutionary; oil peaking will be abrupt and revolutionary."
- The main problems created by Peak Oil will be concentrated on the transportation sector, which relies primarily on petroleum-derived liquid fuels for which there are no readily available substitutes.
- The mitigation efforts needed to avert severe impacts from worldwide Peak Oil could take decades. These efforts will involve replacing "vast numbers of liquid fuel-consuming vehicles" and building "a substantial number of substitute fuel production facilities." The authors state, "There will be no quick fixes. Even crash programs will require more than a decade to yield substantial relief."
- Demand for oil can be reduced through higher oil prices and tighter efficiency requirements. But such reductions will not be sufficient to meet the Peak Oil challenge. Large quantities of substitute fuels will also have to be produced.
- Peak Oil "presents a classic risk management problem" that necessitates careful planning and well-timed mitigation efforts.
- Government intervention will be needed.
- Mitigation efforts are crucial to averting major economic difficulties.

supply disruptions will aggravate price volatility as traders stimulate speculation and as market participants react to changes in supply versus demand. The ultimate effects may be unpredictable. It concludes that "the world has never faced a problem like this. Without massive mitigation more than a decade before the fact, the problem will be pervasive and will not be temporary. Previous energy transitions were gradual and evolutionary. Oil peaking will be "abrupt and revolutionary." A number of major oil company executives are already beginning to publicly acknowledge this fact.

Given the wide range of Peak Oil forecasts and opinions, it comes as no surprise that the battle lines are being sharply drawn. Only a few policymakers appear to appreciate or vocalize the implications of this coming crisis, and fewer still are advocating the need to take immediate action to avoid calamity.

A few policymakers understand that peaking has the potential to affect the entire global food chain, which is currently dependent on cheap fossil fuels, and that following the peak there could be ominous worldwide food shortages; some go so far as to warn that as oil production declines, so must the human population. The Western way of life will be in particular jeopardy because its entire infrastructure and society have been built around relatively abundant, stable, and cheap oil prices.

Meanwhile, those on the other side suggest that the forecasts are propagated by alarmists and that their warnings are simply frightening people. Many on this side of the aisle believe that a yet-to-be-invented technological solution will save the day. Others argue that market forces will provide the answers: As oil becomes more expensive, fewer people will buy it. Higher prices will make alternative energy sources more competitive, and this will fuel development of new policies and alternatives for coping with the peak. As the debate rages over the timing, severity, and implications of Peak Oil, it is becoming increasingly apparent that the era of "cheap" oil is coming to an end.

As it stands now, petroleum prices and policy are being influenced, if not steered by, OPEC. Any comprehensive national energy policy must consider and respond to future decisions that OPEC makes in setting petroleum supplies. As it stands today, the United States and many other Western nations lack a comprehensive scientifically-based policy. Politicians need to begin seriously considering policy options such as the Pickens' Plan.

C.3.1 Energy source policy alternatives

The "old guard" of large-scale energy technologies—coal, oil, gas, nuclear—are unsuitable (at least in the short-term) for solving the peaking

Appendix C: Peak oil: The looming oil crisis

problem, which will largely manifest itself in terms of a liquid-fuels crisis. Over the long term, Peak Oil solutions must be based on sustainable energy policies. Petroleum has been a unique energy source, and we are unlikely to see alternate energy sources developed in the near future that will provide the same advantages:

- Relatively abundant (in the past) and available to most nations with the funds to purchase it
- A massive and expensive infrastructure that has been developed to produce, process, store, and distribute it
- A very high energy density
- Used throughout most of the transportation industry
- Highly versatile energy resource.

Some of the principle energy policy alternatives include the following. As explained below, the key viable energy supply alternatives are limited.

C.3.1.1 Nonconventional petroleum
The world contains large reserves of non-conventional petroleum (oil shale, tar sands etc.) which are as yet largely untapped. Non-conventional petroleum resources may have significant potential for mitigating the effects of Peak Oil; however, economic, geopolitical, technological, and environmental factors may severely hamper the timing and ultimate production rates of these resources.

C.3.1.2 Coal
It's likely that politics and inertia will dictate policies that call for the expanded use of coal in the United States. and in other countries that have plentiful coal reserves, regardless of the significant environmental impacts. It may well be feasible to produce transportation fuels through coal gasification, or coal might be used to produce hydrogen fuel for transportation vehicles. But this also has its own drawbacks (see the hydrogen section below).

C.3.1.3 Hydrogen
Hydrogen is frequently touted as a replacement for petroleum. But, what many fail to realize is that hydrogen is not so much an energy source as it is an energy carrier. Large amounts of energy must be expended to create the hydrogen. More energy must be expended in "transporting" it to stations for use by vehicles. Once created, the hydrogen may provide a relatively clean form of transportation fuel, yet this is not necessarily the case for the process that is used to create the hydrogen in the first place.

According to the U.S. Department of Energy, the transition to a hydrogen economy could take many decades. Significant technical challenges must be solved before the "hydrogen economy" will ever come to fruition. Such a strategy will require development of new and bold national energy policies. Its chief drawbacks are that it has a low energy density, is difficult to store and transport, current hydrogen fuel cells are still quite expensive, and society lacks the enormous existing infrastructure that would be needed to generate and efficiently support a hydrogen economy. Today, 95% of all hydrogen is produced from fossil fuels, which is one of the problems that hydrogen is supposed to solve. Moreover, the inherent inefficiency of using electricity to power an electrolysis process (an electro-chemical process for producing hydrogen) makes it worthwhile only when there is an excess of electrical power generation. As a result of such challenges, hydrogen can't be counted on at this time to mitigate the effects of Peak Oil.[18]

C.3.1.4 Nuclear

In terms of mitigating the coming peak, nuclear energy will provide only a very limited option. The reason for this is that only 3% of U.S electricity is generated from petroleum.[19] Thus, replacing oil with nuclear would only lessen current petroleum usage by an equivalent percentage. While nuclear could play a role in furnishing future energy, it will have little effect in terms of reducing dependency on petroleum. However, it could play a role in terms of producing hydrogen through electrolysis, which would reduce consumption and dependency on petroleum. The largest drawback with such a policy option is that it would take many years to construct sufficient nuclear power plants that could produce the hydrogen. Even with this problem aside, we would still be faced with the problem of having to construct an enormous infrastructure necessary for supporting a hydrogen-based economy.

The Obama administration appears to be increasingly supportive of nuclear energy. Yet, detractors maintain that nuclear power is unsafe. They also argue that nuclear power generates a long-term nuclear waste problem. In contrast, nuclear advocates counter that the potential harm of a large nuclear accident pales in comparison to the worldwide effects of global warming the recent nuclear disaster in Japan has opened this argument to significant criticism.

C.3.1.5 Wind, tidal, solar, and geothermal

One of the most promising sustainable energy transportation technologies may involve wind, solar, and wave/tidal energy that could power a fleet of plug-in hybrids. Wind, tidal, and solar energy technologies have significant long-term electricity-generating potential, which could be used to produce hydrogen. However, for many economic and technological reasons, these technologies are unlikely to fill the gap in the near term. Geothermal power also presents a large potential energy resource that some experts believe is more viable than wind, tidal, or solar power. Policymakers should develop plans for promoting these technologies.

C.3.2 Developing a comprehensive energy policy

Significantly more research is needed immediately to support preparation of any comprehensive national energy policy. Research and development (R&D) funding needs to be given top priority. The authors of the Hirsch report dismiss the power of the markets to solve any oil peak, at least in the short term. They call for rapid government intervention. But in speeding up this decision-making, they also point to a need to limit public debate and environmental analysis. While this may be necessary, it also raises significant issues basic to a democratic way of life.

Any viable policy needs to seriously consider increased tax breaks and incentives to assist homeowners and businesses in installing solar hot water heaters, as well as solar heating and geothermal cooling systems. Although some decent tax breaks exist for homeowners who implement solar hot water and (some) passive solar techniques, few new-home builders offer well-developed options. Thus, as new subdivisions go in, more and more homes add their own unnecessarily high energy loads to the electricity and natural gas grids. If this were done, it would place less burden on electricity and natural gas that could otherwise be used to fuel the transportation sector.

Surviving the peak may require development of innovative new policies that encourage widespread relocalization of agriculture and manufacturing centers to lessen the burden of transporting such products over long distances. Likewise many workers typically commute 50 to 100 miles every day to work. Policies may be needed to encourage workers to live closer to work, work in centers closer to home, or even work at home. Policymakers may also need to consider significant rethinking and expansion of mass transportation systems.

Peak Oil will require public policy choices and many complex and difficult trade-offs. Consider carbon capture and sequestration (CCS), which remains an infant technology. At present there is currently only one demonstrated commercially successful CCS project in the world (the 1,300 MWe coal-fired Mountaineer Power Plant in West Virginia), and it is capturing only a mere 1.5% percent of its emissions. If CCS is ever deployed on a large scale, the CCS technology will consume approximately 30% of the energy it produces to "clean" the coal emissions it produces, and it may add up to 80% in additional production cost. Emphasizing CO_2 emissions in a Peak Oil policy may result in several undesirable outcomes that run counter to the goals of developing an energy mitigation policy. For example, it raises costs substantially, while reducing overall power output; to make up for the power lost in the CCS process, more coal is burned which runs counter to the original purpose of CCS in the first place. This is not to say that we shouldn't be aggressively pursuing carbon emissions reductions, only that the two goals may oppose one another—further complicating the overall problem and tradeoffs have to be made.

C.3.3 Geopolitical repercussions

Much of the world's remaining oil reserves lie in countries unfriendly to the West. As oil supply dwindles, these countries will have increasing influence on the world stage. Wealth will flow into these nations as it flows out of consuming countries. Countries like Iran, the former Soviet Union, and Venezuela with dubious foreign policy motives will gain in wealth, power, and influence, as they hold significant leverage over the consuming nations. It is not difficult for many to envision a future in which some of these nations or perhaps OPEC itself "blackmails" or begins to dictate foreign policy demands to the West. Armed conflict could easily erupt as various groups fight over the control of land, water, food, and energy supplies. We could also witness unprecedented levels of migration, "diasporas," as people migrate or invade countries with more abundant resources. Such migrations would probably be accompanied by violent resistance from those already living in the more desirable areas.[20] This, of course, would significantly raise world hostility and the risk of regional or global conflict. Such concerns need to be seriously considered in any viable national energy policy.

C.3.4 Preparing for the crisis

Surprisingly, during the course of the 2008 U.S. presidential campaigns, not a single major candidate even raised what may be soon come to be one of greatest crises of the last hundred years. As we write this chapter, the authors can point to only one U.S. politician, Representative Roscoe G. Bartlett (Maryland), who has seriously and consistently warned of an escalating risk of a Peak Oil calamity in the future.

Arguably, the best way to prepare for Peak Oil is to consider it in the context of preparing for a natural disaster. The consequences of failing to prepare for Hurricane Katrina were aired live over the television network. In the final analysis, this disaster was the result of failing to prepare for a Category 5 hurricane (which many scientists warned was a distinct possibility). Katrina resulted from a broad failure to prepare for a disaster across a spectrum of local, state, and federal policymaking levels. No comprehensive plan or policy had even been prepared for such a disaster. The result was chaos, confusion, uncoordinated responses, and much suffering.

The consequences of Katrina will pale in comparison to the impacts of Peak Oil. In terms of Peak Oil, decisions being made today regarding purchasing vehicles with higher fuel efficiency, commitment of funding to mass transportation projects, increasing national emergency storage supplies, and carpooling are just a few of the ways that individuals and government policymakers can plan to mitigate the Peak.

C.3.5 A crash program

Some policy platforms (including the Hirsch report) advocate a crash program based on how the United States mobilized to wage World War II following the Great Depression. One of the problems with such an approach is that it does not provide a comparable model. For example, in mobilizing to wage World War II, roughly a quarter of U.S. resources (people and manufacturing capacity), which were idle following the Depression, were brought to bear on this great undertaking. This mobilization stimulated the U.S. economy and moved the nation out of the Great Depression.

In contrast, any crash program to prepare for Peak Oil will operate within a more functioning economy. Allocating resources towards Peak Oil will mean diverting them from other productive areas of the economy, which is sure to cause direct or indirect economic repercussions elsewhere.

If not properly conceived, a crash program could easily "crash." Consider some of the U.S. energy program failures from the 1970s: rationing and price controls (which many economists considered to be a failure) and decades of support for fusion, with little to show for it. Consider just two of many options with significant unknowns and limitations.

1. A rapid shift of vehicle fleets to hybrids requires developing the ability to fuel them on the highway and service them in local repair shops (trained technicians and proper equipment). This simply cannot be done overnight.
2. Building nuclear power plants costs billions of dollars and years of permitting and construction time. It also requires creating university engineering programs to train the nuclear engineers (many have shut down over the last several decades). A shortage of trained engineers may represent one of the largest limiting factors to a goal of significantly expanding nuclear power. Moreover, experienced nuclear talent does not grow on trees; it requires years of professional experience once one has graduated from a university program. Finally, the world's capacity to create specialized nuclear components is limited. There are also serious questions about the supply of uranium fuels.

Added to the complexity of these policy options is a significant body of unknowns. For example, does converting corn to ethanol generate more energy than it consumes? Do we have sufficient coal resources to support a massive clean coal program? Even if we do, how clean and cheap will this energy source really be? Spending money before analyzing alternatives, as we have done with the corn-to-ethanol program, can waste significant time and money.

C.4 Concluding thoughts

The implications of a Peak Oil scenario for our modern industrial society are almost inconceivable. Consider for a moment one single segment—agriculture. In his book, *The End of Food*, Paul Roberts reports that malnutrition was common throughout the 19th century. It was not until the 20th century that cheap fossil fuels allowed agricultural output sufficient to avert famine. It has been widely argued that an exponential increase in energy supply is the principal reason that we have produced a food supply that has grown exponentially in parallel with the increase in human population. Thus, we have avoided wide-scale famine largely because fossil fuel supplies expanded geometrically.

David Pimentel and John Steinhart quantified the energy independence of modern agriculture, and showed that technological development is almost always correlated to an increased use in fossil fuels.[21] The economic expansion of most countries has been shown to be nearly linearly correlated with energy consumption. When this energy increase is constrained, it tends to hamper economic growth. These fossil fuels are used throughout the agricultural cycle (e.g., farm machinery, transportation, pesticides, fertilizers, and drying). Today, it takes approximately 10 calories of petroleum to generate one calorie of food. Nearly 20% of all energy used in the United States is funneled into our food system. If this energy supply begins to tighten and prices escalate, the world could face a nightmare scenario.

As the DOE's Hirsch report makes clear, it is time to seriously and honestly acknowledge the severity of the global Peak Oil emergency. An international program is needed to develop global energy alternatives—something along the lines of the United States' crash program to put a man on the moon in the 1960s. Numerous alternative technologies are already available for implementation. All we need are the requisite funds and the determination to act.

With respect to new technology and governmental action, evidence of the potential impacts has often generally preceded public acceptance and policy actions by several decades. The multifaceted failure of a substantial portion of modern industrial civilization is so completely misunderstood by political leaders that we are virtually unprepared to deal with the outcome. The failure to inject the reality and potential impacts of Peak Oil into the mainstream public policy forum is a grave threat to our modern society.

The important question is not whether Peak Oil will occur, but when it will do so, what the shape of the curve will be, and how steep the slope of the curve will be. The longer we wait, the worse the impact of Peak Oil. We can only imagine how many people will soon be asking, "Why didn't anyone do something about this crisis when we still could?"

Endnotes

1. Pojman, L. P., *Global Environmental Ethics*, p. 284, Mayfield Publishing Company (2000).
2. Miller, G. T., *Living in the Environment*, 10th ed. (Belmont, Calf: Wadsworth, 1998). p. 293.
3. U.S. Crude Oil Field Production (Thousand Barrels per Day), 1859–2006. United States Energy Information Administration. Available online at http://tonto.eia.doe.gov/dnav/pet/hist/mcrfpus2a.htm.
4. A tribute to M. King Hubbert. (2006, February 28). Available online at http://www.mkinghubbert.com/speech/prediction?PHPSESSID=421e8d6f06a74e7afbb37b79f52908ad.
5. Available online at http://www.energybulletin.net/13630.html.
6. See note 1.
7. Grove, N. (1974, June). Oil, the dwindling treasure. *National Geographic*, 145(6), 792–825. This article quoted Dr. Hubbert as stating, "The end of the oil age is in sight."
8. Campbell, C. J. (2003). *The Essence of Oil and Gas Depletion*. Essex, UK: Multi-Science Publishing Company. This book includes estimates of how much oil production can be obtained from conventional sources, as well as from "unconventional" sources such as heavy oil, tar sands, natural gas liquids, and oil shale.
9. de Winter, F. Reading material on the bleak future we can expect in petroleum and natural gas as sources of energy. Available online at http://www.energycrisis.co.uk/Dewinter/.
10. Simkins, C. (2005, September 28). Open letter to Daniel Yergin on optimism and addressing Peak Oil seriously. Energy Bulletin. Available online at http://www.energybulletin.net/9335.html.
11. Oil reserves. (updated 2008, January 11). Wikipedia. Available online at http://en.wikipedia.org/wiki/Oil_reserves.
12. Krauss, C. (2007, December 9). Oil-rich nations use more energy, cutting exports. New York Times. Available online at http://www.nytimes.com/2007/12/09/business/worldbusiness/09oil.html?_r=2&hp&oref=slogin&oref=slogin.
13. Ibid.
14. Hirsch, R. L., Bezdek, R., and Wendling, R. (2005, February). Peaking of world oil production: Impacts, mitigation, and risk management. Available online at http://www.netl.doe.gov/publications/others/pdf/Oil_Peaking_NETL.pdf.
15. Ibid., at p. 4.
16. Ibid., at pp. 64–66.
17. Ibid., at p. 59.

18 National Hydrogen Energy Roadmap, from the results of the National Hydrogen Energy Roadmap Workshop. Washington D.C., April 2–3, 2002. United States Department of Energy. Paper available online at http://www.hydrogen.energy.gov/pdfs/national_h2_roadmap.pdf.
19 *Net Generation by Energy Source by Type of Producer*. EIA.DOE.gov. Energy Information Administration. 2005. Accessed May 2006 at http://www.eia.doe.gov/cneaf/electricity/epa/epat1p1.html.
20 *Resource Wars: An Interview with Michael Klare*. AlterNet. May 1, 2001. Accessed May 2006 at http://www.alternet.org/story/10797/.
21 *American Scientist*, Vol. 97, No. 3, p. 234, May–June 2009.

Appendix D*: Sustainability and energy policy

The terms *sustainability* and *sustainable development* are often used interchangeably. Before proceeding further, however, it is instructive to discuss the difference between these two terms. The term "sustainability" refers to a *long-term* and perhaps unachievable *goal*. In contrast, the term "sustainable development" refers to the more short-term *process* used to *move us closer to the ultimate goal of sustainability*. More to the point, sustainability is the intended goal or outcome, while sustainable development refers to the specific developmental process or steps taken to achieve this goal.[1]

The term sustainability has been applied to everything from local communities, to agriculture and economic systems, to communities and automobiles; on the surface, the concept appears to be relatively simple and straightforward. Yet, despite its almost inherent simplicity, this concept can be deceptively difficult to define, communicate, and implement in practice. As the reader will see, energy management systems (EnMS) can be used to more effectively implement sustainable development policies and proposals.

D.1 Definitions of sustainability

Numerous definitions of sustainability have been proposed that involve adopting a collection of economic, social, and environmental goals that are mutually attainable. The modern use of this term is first attributed to the Brundtland Commission's report, *Our Common Future*, which defined sustainable development as

---

* This chapter is based on Chapter 5 of the author's textbook, Global Environmental Policy: Concepts, Principles, and Practice (Eccleston & March) (2010).

> ... development which meets the needs of the present without compromising the ability of future generations to achieve their needs and aspirations.[2]

The scope of sustainable development can be viewed more comprehensively than by simply considering natural resources. As a more comprehensive concept, sustainability has been defined as

> ... development that delivers basic environmental, social, and economic services to all without threatening the viability of the natural, built, and social systems upon which these services depend.[3]

The International Chamber of Commerce writes:

> ...sustainable development means adopting business strategies and activities that meet the needs of the enterprise and its stakeholders today while protecting, sustaining, and enhancing the human and natural resources that will be needed in the future.[4]

Another definition based more in terms of consumption of goods and services is that of sustainable consumption. According to this definition:

> Sustainable consumption is the consumption of goods and services that have minimal impact upon the environment, are socially equitable and economically viable whilst meeting the basic needs of humans, worldwide. Sustainable consumption targets everyone, across all sectors and all nations, from the individual to governments and multinational conglomerates.[5]

Typically, the concept of sustainable development or sustainability means that the consumptive use of renewable resources does not exceed the regenerative capacity of the environment.[6] Social progress, environmental protection and preservation, conservation of resources, and economic maintenance are all elements of sustainable development. Within these constraints are also quality-of-life concerns, biological and cultural diversity considerations, and conservation, not to mention philosophical questions for humanity. The welfare of future generations also fits into the sustainable development equation.

The concept *sustainable yield* can be thought of as the optimum level of production (e.g., timber, fisheries, water) of a renewable resource that can be maintained indefinitely. In economic terms, it represents the maximum long-term level of income that can be derived from the use of a resource without causing its eventual degradation or depletion.

It is important to note that many ecologists reject the concept of *maximum sustained yield* (MSY) that was promoted in the last century by commercial forestry, agricultural, and fishing interests; they reject this concept because it assumes a long-term stability in the underlying ecosystems that usually cannot be demonstrated to exist. That is to say, natural systems tend to be more complex, more variable, and less stable in response to a disturbance than such a production strategy can account for.

D.1.1 Achieving sustainability

Neil Carter identifies five essential principles of sustainable development:[7]

Equity: Our inability to promote the common interest in sustainable development is often a product of the relative neglect of economic and social justice within and amongst nations ...[8]

Democracy: Sustainable development requires a political system that ensures effective citizen participation in decision-making.[9]

The Precautionary Principle: In order to protect the environment, the precautionary approach should be widely applied by states according to their capabilities. Where there are threats of serious or irreversible damage, lack of full scientific certainty will not be used as a reason for postponing cost-effective measures to prevent environmental degradation.[10]

Policy Integration: The objective of sustainable development and the integrated nature of global environmental/development challenges pose problems for institutions... that were established on the basis of narrow preoccupations and compartmentalized concerns.[11]

Planning: Sustainable development must be planned.

Moreover, Principle 4 of the United Nations *Agenda 21* declares:

> In order to achieve sustainable development, environmental protection shall constitute an integral part of the development process and cannot be considered in isolation from it.

The concept of sustainable development is also one of the pillars of the United Nation's eight *Millennium Goals* pursuant to its population policy. Goal #7 states:

> Goal 7: Integrate the principles of sustainable development into country policies and programs and reverse the loss of environmental resources.

Specifically, Goal #7 declares the following "targets" pursuant to that goal:

1. Integrate the principles of sustainable development into country policies and programs, and reverse the loss of environmental resources.
2. Reduce biodiversity loss, achieving, by 2010, a significant reduction in the rate of loss.
3. Halve, by 2015, the proportion of the population without sustainable access to safe drinking water and basic sanitation.
4. By 2020, to have achieved a significant improvement in the lives of at least 100 million slum dwellers.

D.2 Sustainability impact

Corporations, organizations, and government agencies have all reported numerous "sustainability success stories." But in reality, many of these success stories have done little more than reduce the rate of unsustainability. There is a significant difference between "sustainability" and "reducing the rate of unsustainability"—one that is rarely communicated. MacLean notes that in his experience, most sustainability managers hired by large companies have limited backgrounds in environmental science and regulation, and in fact were extensively trained in communication, marketing, and product development.[12] He concludes that they were principally hired to increase profits and build the brand name recognition, with the goal of achieving truly sustainable products relegated to a distant priority.

Consequently, marketing green products does not necessarily translate to a measurable environmental benefit or a commitment to sustainable development. Packaging an environmentally destructive product in a biodegradable wrapper may appeal to environmentally conscious consumers, but it is doing next to nothing for the environmental health of the planet.

In diverting discussion of sustainable development away from environmental compliance, many corporations have been able to both change public opinion and successfully market questionable "environmentally friendly" products. The problem is that both the public and companies may begin to believe their headlines and ignore the tough choices that must be made to achieve environmental stewardship.

D.2.1 Measuring impact: A rough approximation

One of the key precepts of sustainable development advocates is the concept that environmental effects can be mitigated with advances in technology despite increases in the number of consumers and their per capita income. This school of thought holds that the human impact on the ecosystems can conceptually be expressed and compared by an *Impact, Population, Affluence, and Technology (IPAT)* index:

$$\text{Impact} = \text{Population} \times \text{Affluence} \times \text{Technology}$$

Population and technology play an important role in terms of sustainable development. However, others argue that the role of affluence in this formula plays a much larger role than consumption patterns would indicate, and is more difficult to describe and quantify. Moreover, increased wealth may overwhelm the ability of technology to mitigate the environmental impact of increased population.

D.2.2 Technological paradox

Many optimists believe increased technology will more than offset the resource depletion that results from affluence through the invention of substitutes, discovery of new replacement materials, or improved efficiencies in reuse and recycling. Critics counter that the lure of sustainability has been obscured by the promises that technological advances will yield increased affluence forever.

Over the last few generations, resource prices (the ultimate measure of supply and demand) have generally decreased with time, while the standard of living has improved. Paradoxically, however, lower cost and improved technologies have frequently reduced recycling and conservation measures, causing accelerated resource depletion.

Although modern technology has moderated prices and made resources more abundant, at least in the short term, the number of proven reserves has often not kept pace. This observation applies to many nonrenewable ore deposits, as well as renewable resources such as fresh water, top soil, rain forests, and fish populations.

Michael and Joyce Huesemann analyzed the question "Will technological progress avert or accelerate global collapse?" They came to the conclusion that technology alone is insufficient to achieve sustainable development. While technological progress has improved efficiencies, it has also increased both the number of consumers and their affluence. Without a major environmental policy change, they concluded that technological progress will ultimately cause an impending collapse.

Current projections indicate that the earth's population will stabilize at around 9 billion by the middle of the 21st century. The implications of this global rise in population is that individual consumption levels resulting from increased affluence are likely to grow disproportionately by a much larger amount.

D.3 Agenda 21

The concept of sustainability became the baseline goal for *Agenda 21*, the 40-chapter document that details the goals and programs resulting from the United Nations Conference on Environment and Development (informally known as the Earth Summit), held in Rio de Janeiro, Brazil, in June 1992. Agenda 21 provides 27 principles for implementing its strategy (see Table D.1). Nearly half of these sustainable development principles focus on actions undertaken by national governments. The remainder focuses on actions undertaken by individuals and organizations.[13]

The first principle declares: Human beings are at the centre of concerns for sustainable development. They are entitled to a healthy and productive life in harmony with nature. Pursuant to these principles, the UN promulgated a 40-chapter implementation plan. Seven of the areas covered by this plan explicitly implement the sustainability principle.

In the year 2000, with Agenda 21 in mind, the UN identified eight Millennium Development Goals. In 2002, the UN World Summit on Sustainable Development held in Johannesburg, South Africa, benchmarked the world against the goals and agenda, generating an improved implementation plan.

The concept of sustainable development is also embedded within ISO 14001, an increasingly popular international standard for environmental management systems. The ISO 14001 standard embraces Agenda 21 from the Earth Summit along with strategies of the International Chamber of Commerce business charter for sustainable development.

D.4 Common principles of sustainability

Although the U.S. National Environmental Policy Act (NEPA) predates the modern concept of sustainable development, the rudimentary concept

Appendix D: Sustainability and energy policy

Table D.1 Summary of the Twenty-Seven Principles Contained in Agenda 21

1. Human beings are entitled to a healthy and productive life in harmony with nature.
2. States have the right to exploit their own resources but without damage to others.
3. The right to development must meet the needs of present and future generations.
4. Environmental protection is an integral part of the development process.
5. People must eradicate poverty to decrease disparities in standards of living.
6. Needs of the least developed and most environmentally vulnerable must be a priority.
7. States must cooperate to conserve, protect, and restore Earth's ecosystem.
8. States are to eliminate unsustainable patterns of production and consumption.
9. States are to improve scientific understanding to strengthen capacity-building.
10. Environmental issues are best handled with participation by all concerned.
11. States must enact effective environmental legislation.
12. States are to promote supportive, open economics for growth and development.
13. States must have laws to protect victims of pollution and environmental damage.
14. States are to cooperate to discourage and prevent severe environmental degradation.
15. The precautionary approach must be applied to threats involving serious damage.
16. Authorities are to promote "polluter pays" with due regard to the public interest.
17. Impact assessments must be performed to evaluate likely significant adverse impacts.
18. States must notify others of disasters or emergencies likely to harm others.
19. States must notify others of transboundary environmental effects.
20. Women's full participation is essential to the goal of achieving sustainable development.
21. World youth partnership is essential to achieve sustainable development.
22. Indigenous people and communities have a vital role in sustainable development.
23. The environment and resources of people under oppression are to be protected.
24. Warfare is inherently destructive of sustainable development.
25. Peace, development, and environmental protection are interdependent and indivisible.
26. States must resolve environmental disputes peacefully.
27. States and people must partner to meet the goal of sustainable development.

is nevertheless embedded in the act. Consider the following two excerpts from the environmental policy that NEPA founded:

> ...productive and enjoyable harmony between man and his environment; to promote efforts which will prevent or eliminate damage to the environment and biosphere and stimulate the health and welfare of man...[14](Emphasis added).
>
> ...it is the continuing policy of the Federal Government to use all practicable means and measures, including financial and technical assistance, in a manner calculated to foster and promote the general welfare, to create and maintain conditions under which man and nature can exist in productive harmony, and fulfill the social, economic, and other requirements of present and future generations of Americans.[15] (Emphasis added).

The U.S. Environmental Protection Agency (EPA) has taken its own path toward describing sustainable development. The EPA's Center for Sustainability promotes linking environmental, economic, and social goals to "enhance" quality of life and encourage livable communities that can someday realize a "New American Dream." It recommends protecting vital resource lands, conserving energy and nonrenewable resources, and reversing unsustainable transportation trends.

The concept of sustainability has also received significant international attention, particularly within Europe and among other industrialized nations. For example, in 1991, the *Resource Management Act of New Zealand* was enacted. This act set a new precedent by articulating what some experts have called the world's first legislative statement promoting the principle of sustainability.

D.4.1 U.S. executive order on sustainability

A recent U.S. Executive Order (E.O.), E.O. 13514, *Federal Leadership in Environmental, Energy, and Economic Performance,* directs federal agencies to establish an integrated strategy for sustainability and make reduction of GHG emissions a federal agency priority.[16] This E.O. directs agencies to enhance other aspects of sustainability by reducing water consumption, minimizing waste, supporting sustainable communities, and using federal purchasing power to promote environmentally responsible products and technologies.

Under this E.O., each agency must submit an annual Strategic Sustainability Performance Plan, subject to approval by the OMB Director, to address, among other topics:

- Sustainability policy/goals, including GHG reduction targets
- Integration with agency strategic planning/budgeting
- Schedules/milestones for activities covered by the E.O.
- Evaluation of past performance based on net lifecycle benefits
- Planning for adaptation to potential climate change

D.4.2 Principles can be found in most pragmatic sustainable development

Hargroves and Smith (2005) have identified a number of common principles that can be found in most pragmatic sustainable development programs.[17] These principles are depicted in Table D.2.

D.4.3 Adoption of sustainability policies

The goal of sustainability requires that a proactive approach encompass economic development and preservation of environmental quality. The overlapping roles of governance with ecological, social, and economic needs generate the interdisciplinary nature of sustainable science. How we do this depends partly on our perceptions and on how we frame the environmental issues. If policymakers perceive resource depletion as the prevailing issue, they may emphasize recycling policies; if they believe that there is an unlimited amount of extractable resources, or new technologies and substitutions are on the horizon, they may

Table D.2 Common Principles can be Found in Most Pragmatic Sustainable Development Programs

Integration of economic, social, and environmental goals in policy formulations
Commitment to best practices of sustainable development
Promoting continuous improvement
Dealing transparently and systemically with risk and uncertainty
Ensuring appropriate valuation, appreciation, and restoration of natural resources and environs
Conserving biodiversity and ecological integrity
Avoiding the loss of human capital as well as natural capital
Providing opportunities for community or stakeholder participation
Being cognizant of intergenerational equity

have little incentive to protect such resources. Similarly, if the public and politicians believe that automobile emissions are causing global warming, they may promote more fuel-efficient cars, or seek other alternatives.

Moreover, citizens must feel empowered and must believe that their efforts are meaningful. Individual perceptions mold the collective will, and hence shape policies that can institutionalize what may have begun as simple lifestyle choices. With respect to automobiles, monetary incentives can encourage people to do the right thing (purchase more fuel-efficient vehicles). Conversely, prohibitions are more heavy-handed policy instruments. In forging a sustainable development policy, the beliefs of key policymakers, the system of governance, and the role of individuals and organizations all play a role.

D.4.4 Sustainability hierarchy

Marshall and Toffel first proposed a *Sustainability Hierarchy* to provide a framework around a variety of issues that have been associated with sustainability.[18] This hierarchical system categorizes actions as unsustainable based on their direct or indirect potential to:

1. Endanger the survival of humans
2. Impair human health
3. Cause species extinction or violate human rights
4. Reduce quality of life or have consequences that are inconsistent with other values, beliefs, or aesthetic preferences.

Marshall and Toffel argue that for sustainability to become a more meaningful concept, the various worthy issues in the fourth category (values, beliefs, and aesthetic preferences) should not be considered sustainability concerns.

D.4.4.1 The economic costs of sustainability

The escalating costs of fossil energy and material commodities have much to do with motivating environmentally sustainable policies. Arguably, the principle reason U.S. automakers are currently in economic trouble is that they have pushed high-profit, gas-guzzling SUVs and trucks to the detriment of more fuel-efficient vehicles. When the price of crude oil skyrocketed in 2008, the market for big vehicles began to crash. Unfortunately, they had no one to blame for this situation but themselves. Over the long run, rising energy prices will stimulate research into more fuel-efficient and sustainable technologies.

Frequently, it does not require radical new technologies but merely a new way of looking at an existing business model. The biggest challenge to marketing electric cars has involved the development of a new generation of batteries that can power cars over a longer—or perhaps even unlimited—range. Clearly, range limitation is a significant reason many people shy away from electric vehicles. Price is another factor. But what if the price and range problem can be overcome without developing a new generation of electric batteries?

Shai Agassi, an Israeli entrepreneur, wants to do just that—electrify the world's car fleet—not by developing a new battery that gets 250 miles on a charge, but by providing a system that allows drivers to travel as far as they want without stopping for a time-consuming recharging. He proposes to establish a system of highway battery exchange stations, analogous to the gas canister exchanges that people now use for barbecue grills. Drivers would simply pull into a station, and the drained battery would be swapped for a fully charged one within a few minutes.[19] Innovative ideas like this lie at the heart of sustainable development.

D.4.5 Sustainable dilemmas

What might at first appear to be a sustainable policy can on closer inspection be anything but. Policies that are promoting production of fuel ethanol from corn are a case in point. Corn is a renewable resource; in fact it's a solar-based product. What sounds cleaner and more sustainable than a national ethanol program?

It's a great idea until one performs a more thorough life-cycle analysis; studies of the U.S. ethanol program reveal the hidden high energy-intensity practices of cultivating and harvesting corn and its conversion into ethanol fuel. An ethanol program is only marginally more efficient and sustainable than producing gasoline.

But the story does not end here. Diversion of corn to manufacture ethanol can only mean less corn left to feed livestock and people, which drives the cost of food up. In some cases, rainforests in locations such as Brazil are being destroyed to produce farmland to grow crops to produce fuel in the United States. The loss of valuable rainforests in turn increases carbon dioxide concentrations, which negates one of the most powerful arguments for manufacturing ethanol in the first place. We cannot confidently declare any practice as "sustainable" until a full life-cycle analysis of its environmental costs has been performed.

D.5 Measuring sustainable development

It is commonly recognized that sustainable development does not focus entirely on the environment or energy. John Elkington coined the term *triple bottom line*, which refers to three very broad criteria to measure organizational or societal success: Economy, Society, and Environment. [20] Environmental policy analysts have long relied on economic models that seek to convert physical environmental impacts on the air, water, land, and biota into surrogate marketplace dollars. Many critics of dollar measures for nonmarket environmental effects have proposed alternative indices to guide environmental decision-making. In this section, we describe two significant efforts to develop quantitative tools that implement this idea.

D.5.1 The ecological footprint (EF)

The following section describes two methods in use for assessing ecological footprints.[21]

D.5.1.1 The global footprint network

Mathis Wackernagel and William Rees of the University of British Colombia developed the concept of an "ecological footprint." Since that time the Global Footprint Network (GFN), an organization that promotes sustainable development, has matured the concept into a standard methodology that has been adopted by many public and private sector organizations as a research and decision tool. The term is defined as follows:

> **Ecological Footprint:** A measure of how much biologically productive land and water an individual, population, or activity requires to produce all the resources it consumes and to absorb the waste it generates using prevailing technology and resource management practices. The Ecological Footprint is usually measured in global hectares.

One concept, the *ecological footprint*, provides a measure of the impact of different kinds of human activities that use air, water, land, and biotic products that support a given lifestyle tied to a level of technology and economic activity. GFN has tabulated generic per/capita footprint for regions, cities, and corporate enterprises. The EF enables those who analyze proposed policies and development plans to compare the EF values of any number of alternative development schemes. When incorporated into environmental impact assessment reports, they are helpful in choosing the best option among the alternatives.

> **Biocapacity (Shorthand for biological capacity):** The ability of an ecosystem to produce useful biological materials and to absorb wastes generated by humans.

Appendix D: Sustainability and energy policy

For an ecosystem to sustainably meet human needs (food, fiber, energy, minerals, water, etc.), its biocapacity measured in hectares must equal or exceed its rate of consumption by a given population. The result of the calculation is defined as follows.

D.5.1.2 Ecological rucksack[22]

An alternative concept referred to as the *Ecological Rucksack* provides another index approach that has been used as an environmental policy metric. The ecological rucksack is defined as the total quantity (typically in kilograms) of materials moved from nature to create a product or service, minus the actual weight of the product. That is, an ecological rucksack resembles a hidden material flow assessment. Ecological rucksacks take a life-cycle approach and signify the environmental strain or resource efficiency of the product/service manufactured or produced. More specifically, ecological rucksacks measure the amount of materials not directly used in the product but displaced because of the product. That is, ecological rucksacks represent the materials necessary to produce, use, recycle, and dispose of a product, but not the materials specifically used in the product.

The ecological rucksacks use a cradle-to-grave approach. They focus attention on the entire life cycle of a product or service and the environmental and resource impacts that are consumed in producing it.

D.5.1.2.1 Simplified methodology Friedrich Schmidt-Bleek from the Wuppertal Institute for Climate, Environment and Energy (Germany) first proposed the ecological rucksack concept. The ecological rucksack (ER) is calculated by subtracting the weight (W) of the product from the material intensity (MI) of the product or service as follows:

$$ER = MI - W$$

The material intensity is found using:

$$MI = SUM (Mi \times Ri)$$

Where:
  Mi is weight of a material in kilograms (kg)
  Ri is the rucksack factor.

The material input is calculated from five main categories:

1. Abiotic raw materials
2. Biotic raw materials

3. Moved soil (agriculture and forestry)
4. Water
5. Air

The rucksack factor is the quantity (in kg) of materials moved from nature to create 1 kg of the resource. For example, the rucksack factor for aluminum is 85:1 (85 kg of materials moved for every 1 kg of aluminum produced), and for recycled aluminum it is 3.5:1. For a diamond, the rucksack factor is 53,000,000:1.

Each material used in the production of the good or service is multiplied by its rucksack factor, and then each normalized value is summed to produce the material intensity of that product or service. As far as possible, all materials used for the production, use, and disposal, whether directly or indirectly, are included in the calculation.

The ecological rucksack of some materials will change over time as they become rarer or as technology makes extraction or processing more efficient. For example, copper has moved from an ecological rucksack of 1:1, when copper was easier to find, to 500:1 as most copper is now being extracted from sulfide ores.

Case Studies Examples

1. ECOLOGICAL RUCKSACK VALUES

The Association of Cities and Regions for Recycling provides some rucksack values for various products. A 5 gram gold ring, for example, was found to have an ecological rucksack of 2,000 kg. An aluminum drink can was found to have an ecological rucksack of 1.2 kg. And a 20 kg computer was found to have an ecological rucksack of 1,500 kg.

2. REDUCING THE ECOLOGICAL RUCKSACK OF A WATCH

By conducting a Material Intensity Per Unit Service (MIPS) Analysis for the production phase of a new watch, opportunities to reduce the ecological rucksack of the watch were revealed. By changing some materials used in the watch, the rucksack decreased.

D.5.2 Policy options for reducing the impact or footprint

Both the Environmental Footprint and Rucksack take into account the relationship between a population's use of natural resources and the ability of the environment to sustainably supply resources. Whichever method is used, controlling population is arguably the most direct policy means of reducing the environmental impact. There are generally two options available: (1) reduce population fertility and (2) reduce immigration (legal and illegal). However, the following policy options can help meet society's needs while reducing or eliminating the need for population controls.

D.5.2.1 Increasing the carrying capacity with science and technology

This has been summarized as increasing "the size of the pie." Efficiency could play a pivotal role in increasing carrying capacity and therefore the abundance of resources. The Cornucopians (growth optimists) would argue that science and technology innovations can be tapped to increase food yields, recycle water, find alternatives to scarce materials, and a host of other remedies; instead of reducing population, many (but certainly not all and perhaps not even a majority) of them would argue that population should remain unconstrained. After all the more people there are, the more scientists, engineers, and entrepreneurs are available for finding solutions to pressing environmental resource problems. Within this group may lie the next Einstein who will find the "silver bullet" for resolving Earth's mounting problems.

D.5.2.2 Reduction in lifestyle or living simpler lives

Limit conspicuous consumption. Reduce our dependency on material and energy-intensive gadgets and devices. Ride bikes instead of cars. Incorporate recycling and conservation into our daily lives.

D.5.2.3 Reforming and redistributing resources

Assist poor countries in developing green technology. Provide food, educational programs, and technological assistance to developing countries. Compensate developing countries for preserving their rainforests.

D.6 Applications of sustainable resource development

The following section provides two contrasting examples of how the sustainability principle has been applied to specific global and national issues. The first provides a global perspective on a sustainable global economy. The second is a business view of sustainable development.

D.6.1 Application 1: Business and Dow Jones sustainability concepts

The World Business Council for Sustainable Development is a coalition of 180 companies from some 35 countries. Each member shares the commitment to sustainable development via three pillars: economic growth, ecological balance, and social progress.

Consistent with this commitment, the Dow Jones Corporation, known for its business and financial indices and other publications, measures sustainability in three dimensions: economic, environmental, and social responsibility (Table D.3). It publishes a family of indexes to track performance by companies in terms of corporate sustainability as defined by the Dow Jones measures.

Table D.3 Dow Jones Measures of Sustainability

Environment
1. Policy/Management
2. Performance (Eco-Efficiency)
3. Reporting (Content & Coverage)

Economic
1. Codes of Conduct, Compliance, Corruption, and Bribery
2. Corporate Governance
3. Customer Relationship Management
4. Investor Relations
5. Risk and Crisis Management
6. Brand/Supply Chain/Marketing Practices Criteria
7. Innovation/Research and Development/Renewable Energy Criteria

Social
1. Citizenship/Philanthropy
2. Stakeholder Engagement
3. Labor Practice
4. Human Capital Development
5. Social Reporting
6. Talent Attraction and Retention
7. Product Quality/Recall Management
8. Global Sourcing
9. Occupational Health/Safety
10. Healthy Living
11. Bioethic

Appendix D: Sustainability and energy policy

The Dow Jones concept of corporate sustainability is a business approach, designed to:[23]

> ... create long-term shareholder value by embracing opportunities and managing risks derived from economic, environmental and social developments. Corporate sustainability leaders harness the market's potential for sustainable products and services while at the same time successfully reducing and avoiding sustainable costs and risks.

D.6.2 Application 2: Worldwatch innovations for a sustainable economy

A report by the Worldwatch Institute, *State of the World: Innovations for a Sustainable Economy,* begins by citing the following global phenomena that raise questions about the continued sustainability of the human economy given the scale of its environmental impacts on vital natural resources essential to that economy. According to this report, we could be facing:[24]

- The possible collapse of currently fished species.
- The rapid increase of oxygen-depleted dead zones in the oceans from 149 to 200 in the past two years, posing threats to fish species.
- Urban air pollution estimated to cause 2 million premature deaths each year.
- Declines in the populations of bees, bats, and other vital pollinators across North America, threatening crops and ecosystems.
- A potential peaking of the world's oil production (see Appendix C).

Actions promoted in these and other sections of Worldwatch come together in Chapter 13 of the report, "Investing in Sustainability," summarized here as follows.[25] Specifically, investment policies are divided into three categories:

Socially Responsible Investments that focus on environmental and social sustainability, and include investment in certain values not directly related to sustainability:

Project Finance: Funding for major infrastructure or extractive projects such as dams and mines that add value to social and environmental sustainability

Private Equity and Venture Capital: The focus here would be on green energy and products

Microfinance: Involves very small loans widely distributed to small-scale artisans and craftsmen to serve local markets. The objective is to increase income among the poorest populations and alleviate poverty

All of these strategies are actually pursued by various United Nations environmental and social programs, and by a range of international donor organizations, often funded by major corporations, that select specific issues or areas of geographic focus for their scope.

D.7 Examples of a sustainable energy policy

The maximum worldwide power consumed at any given moment is about 12.5 trillion watts (terawatts, or TW). The U.S. Energy Information Administration projects that in 2030, the world will require 16.9 TW. Is it possible to develop a global, clean, and sustainable energy system? This section investigates one plan for achieving a sustainable energy infrastructure within two decades using only proven technologies or those that are close to working today on a large scale.

Three dramatically different approaches for achieving a national sustainable energy policy in the following sections. The first plan provides a comprehensive approach in the way that energy is both generated and used; it has been characterized by many critics as more hype than reality. Still, it is useful for illustrating the rudiments of sustainable development, such as a national energy sustainability plan. The second plan provides a different and radically innovative approach for generating and allocating energy, principally for use in the transportation industry. The purpose of describing these plans is not so much to promote them as to provide potential examples of the types of sustainable development policies that can be formulated and implemented. Potential sustainable development policies could be evaluated and implemented using a strategic EnMS for the overarching proposal in conjunction with local EnMSs for implementing specific projects. The strategic EnMS provides overall direction for the proposal, while project-specific EnMSs would evaluate, implement, and monitor specific details of the project.

D.7.1 Plan 1: Combination power plan

Members of *Discover* magazine teamed up with National Science Foundation, Institute of Electrical and Electronics Engineers, and the American Society of Mechanical Engineers to sponsor a series of briefings on Capitol Hill in which they promoted a combination power plan. The plan consisted of three key strategies.[26]

Appendix D: Sustainability and energy policy

1. *Energy Efficiency*: The first component involves increased energy efficiency, which if implemented correctly, is the equivalent of developing a major new source of energy. For instance, properly constructed or retrofitted buildings offer enormous potential for energy savings. Natural daylight can replace 30%–60% of electrical lighting, natural ventilation can reduce electrical air-conditioning consumption by 20%–40%, and natural shading can account for another 10% reduction in energy, while passive heating can eliminate 20%–40% in heating costs. Improved energy efficiency could save 50%–75% on the cost of running equipment and appliances.[27]
    Stronger building codes or retrofitted buildings (improved insulation and lighting) could reduce building energy consumption by 6%–7% by 2030.[28]
    Labeling on consumer products such as the U.S. EnergyStar program can inform consumers about the energy consumption and cost of operating equipment and appliances.
2. *Smart Grid and Improved Energy Storage Systems*: A *Smart Grid* could significantly improve energy efficiency through the use of "intelligent" digital devices that monitor and efficiently allocate electrical demand and supply. For instance, certain equipment and appliances can be programmed to turn off during peak electrical load and to be activated during periods of low load. As another example, improved battery and storage devices could store energy from wind farms during periods of low demand and release it to the grid during periods of peak load.
3. *Alternative Energy*: Oil, natural gas, and coal make up 84% of the total U.S. energy consumption.[29] Catalysts have been discovered that can turn biomass into fuel in a matter of minutes. For instance, laboratory tests have made dramatic advances in producing advanced biofuels (from wood chips and cornstocks). New techniques for efficiently splitting water into oxygen and hydrogen are under study.

However, substantial investment is needed to move this technology from the laboratory to the pilot and demonstration plant phase. Increased funding for wind farms, biofuels, hydrogen fuel cells, and solar could provide significant long-term payback. New regulations, loan guarantees, and cost-sharing programs are also needed.

An immediate national priority needs to be assigned to an Energy Research and Development (R&D) Program. U.S. spending on R&D has hit a 30-year low. The pet food industry spends more R&D capital (in percentage terms) than the U.S. spends on energy. Spending on energy R&D by the U.S. Department of Energy has plummeted from 6 billion in 1978 to 1.8 billion in 2004.

The three aforementioned strategies will require graduating thousands of additional engineers trained in scientists and electrical power. Large amount energy and of R&D should be directed at such things as developing the smart grid, alternative energy, and green buildings.

This three-tiered strategy may have the potential to revolutionize solutions for addressing the daunting energy problems that we are faced with over the next several decades. The Pickens Plan described in the following section provides a very different yet equally promising approach to solving the energy escalating dilemma.

D.7.2 Plan 2: The Pickens sustainable energy plan

We now consider an alternative sustainable energy policy that is less comprehensive in scope than the one portrayed earlier. However, the authors of this book believe that the Pickens Plan is equally pragmatic and subject to less uncertainty than the aforementioned proposal.

In recent years, wind power has experienced near exponential growth in the United States. Oil tycoon T. Boone Pickens has proposed a U.S. national sustainable energy policy, widely referred to as the *Pickens Plan*.[30,31] He specifically wants to strengthen U.S. national energy security by exploiting domestic energy sources while decreasing dependence on vulnerable imported foreign oil. Import of foreign petroleum and natural gas has caused a significant negative trade deficit that is decreasing U.S. national wealth. According to one estimate, the Pickens Plan could reduce the amount the U.S. spends annually on foreign oil imports by $300 billion annually (a 43% decrease).

D.7.2.1 The plan

Pickens wants to reduce American dependence on imported oil by investing $1 trillion in enormous renewable energy wind turbine farms. This would allow natural gas currently burned in electrical power plants to be shifted to fuel trucks and heavy vehicles. Specifically, his plan would involve:

1. Private industry funding and construction of thousands of wind turbines in the Great Plains region of the United States. (i.e., the U.S. "wind corridor" because of its favorable wind patterns). He estimates that these turbines could provide 22% or more of the nation's electricity supply.
2. The U.S. government would fund and construct electric power transmission lines to connect the turbine farms to the nation's power grid. This would provide electrical energy to the Midwest, South, and Western regions of the Unites States. The new transmission lines would cost between $64 billion to $128 billion.[32] Testifying before the

U.S. Senate Homeland Security and Government Affairs Committee, Pickens said that the transmission lines should have the same status priority that President Eisenhower assigned in declaring priority emergency to build the interstate highway system in the 1950s and 1960s.

3. With wind energy providing a large portion of the nation's electricity, natural gas currently used to fuel electrical power plants would be diverted and used as fuel for thousands of trucks and heavy vehicles. This would substantially reduce oil (diesel) use to power trucks. The plan places emphasis on developing a national fleet of trucks and buses that would burn this relatively clean natural-gas resource. This would also reduce air emissions and other environmental impacts.

According to Pickens, the plan could provide 22% of the nation's electrical power supply from wind power and would provide for the conversion of vehicles from gasoline to natural gas in less than 10 years. The plan calls for increasing the installed wind power capacity by at least a factor of ten by 2018 from its 2008 level.

The Pickens plan would significantly reduce CO_2 emissions by shifting a large percentage of electricity production from natural gas combustion to carbon-neutral wind generation. While the natural-gas-powered vehicles would still produce CO_2, they would produce about 25% less when compared to the same amount of energy derived from gasoline/diesel.[33]

Pickens does not believe that electric batteries are likely to power the nation's large trucks, which is why his plan focuses on diverting natural gas to power large vehicles. However, the wind-generated electricity could be used to power electric-battery-powered automobiles.

D.7.2.2 The intermittency problem

A key challenge is that the wind energy used to replace natural gas does not blow some of the time. But this may not pose a significant obstacle. Modern technology, including digital control systems, could allow multiple wind farms from varying regions of the country to be connected together on the electrical grid. While some are idle, others could make up the difference.[34]

Others have argued that to produce power when the wind is not blowing, backup natural-gas-powered plants may be needed. These backup plants could be brought online to help supply peak capacity during periods of low wind or peak demand.

Another suggestion is that instead of producing additional electricity, energy could be conserved. For instance, the Smart Power Grid (See Appendix A), an intelligent electrical distribution network under development, could reduce power consumption during peak hours and thereby lessen dependence on gas-fired plants. Still another option, one which would be more expensive, involves storing energy by using methods such as pumping water uphill when the wind is blowing, and releasing it through turbine generators when the electricity is needed.

D.7.2.3 Cost

According to a study performed by the Cato Institute, the wind energy component would result in higher retail electric rates because wind power is twice as expensive as natural-gas-fired generation. However, the American Wind Energy Association counters that the cost of wind power has dropped by 90% over the last 20 years, and further declines can be expected. Also, the cost of economic "externalities" of gas/diesel powered vehicles is greater that natural-gas-powered ones.

D.7.2.4 Endorsements and criticism

Barack Obama has publicly stated he supports many elements of Pickens Plan. Representatives of the Sierra Club and Center for American Progress have also endorsed his plan.

In contrast, the Institute for Energy Research, an organization funded by the oil industry that advocates off-shore drilling, claims that the Pickens Plan relies on government subsidies and that producing large amounts of wind power is not a viable option.[35] Many disagree with this assessment.

D.7.3 Plan 3: Jacobson–Delucci sustainable energy plan

In developing their energy plan, Jacobson and DeLucci[36] assumed that most fossil fuel heating (e.g., ovens) can be replaced by electric systems, and that most fossil fuel transportation can be replaced by battery and fuel-cell vehicles. Collectively, wind, water, or sunlight are referred to as WWS. The renewable technologies considered in this plan involve low generation of greenhouse and air pollutant emissions over their entire lifecycle (an assertion that has been disputed by some critics), including mining, milling, construction, operation, and decommissioning. Electrolysis (powered using WWS electricity) would produce hydrogen for use in fuel cells. The hydrogen would also be burned in airplanes and used by industry.

They chose a mix of different technologies, emphasizing wind and solar, with nearly 10% produced using mature water-related methods.[37] Wind provided by 3.8 million large wind turbines (each rated at 5 MW) would supply about 50% of the worldwide demand. An additional 40% of the power would come from photovoltaics and solar power plants, with 30% of the photovoltaic output from rooftop panels on homes and buildings. Nearly 90,000 photovoltaic and concentrated solar power plants, averaging 300 MW, would need to be constructed. The plan also includes 900 hydroelectric stations worldwide, 70% of which have already been constructed.

Compared with fossil fuels, electrification tends to be a more efficient way to use energy. For instance, only about 17% to 20% of gasoline energy is actually used to move a vehicle (the rest is wasted as heat). In comparison, 75% to 85% of the electricity delivered to an electric vehicle is used to move it.

D.7.3.1 Materials scarcity

One of the most problematic materials required for such a mammoth undertaking involves rare earth metals such as neodymium used in turbine gearboxes.

Photovoltaic cells rely on materials such as amorphous or crystalline silicon and cadmium telluride. Limited supplies of materials such as indium could reduce the prospects for some types of thin-film solar cells. Recycling old cells might help ameliorate material difficulties.

Components for building millions of electric vehicles include rare earth metals for electric motors, lithium for lithium ion batteries, and platinum for fuel cells, which are scarce resources. Again, recycling could help ameliorate some of the material challenges.

D.7.3.2 Cost

The authors of this plan claim that their mix of WWS sources can reliably supply a nation's entire energy demand. According to their plan, the cost of wind, geothermal, and hydroelectric are all less than 7 cents a kilowatt-hour (¢/kWh). Wave and solar are higher, but by 2020 and beyond, wind, wave, and hydro are expected to be equal to or less than 4¢/kWh.

In contrast, the authors claim that the average cost of U.S. conventional energy transmission was about 7¢/kWh, and it is projected to rise to 8¢/kWh by 2020. In contrast, wind turbine energy already costs about the same or less than it does from a new coal or natural gas plant.

Today, solar power is still relatively expensive, but is expected to become competitive as early as 2020. One study concluded that within 10 years, photovoltaic system costs could drop to about 10¢/kWh, including transmission and the cost of compressed-air storage of power for use at night.

WWS construction cost over a 20-year period might run on the order of $100 trillion worldwide, not including transmission costs. Each nation would need to invest in a robust, long-distance transmission system that could carry WWS power from remote regions where it is often most abundant (such as the U.S. Great Plains for wind and the desert Southwest for solar) to major population centers.

D.7.3.3 Reliability

The authors claim that WWS technologies typically have less "downtime" than traditional power sources. The average offline time for a coal plant, for example, is 12.5% of the year for maintenance. In comparison, wind turbines have a downtime of less than 2% and 5% at sea. Photovoltaic systems average about 2%.

The principal WWS challenge is that the wind does not always blow and the sun does not always shine. However, the WWS authors argue that the intermittency problems can be mitigated by balancing the power sources. For example, base electricity supply could be generated from steady geothermal or tidal power, while relying on wind at night (when it tends to be more plentiful) and using solar by day, and utilizing hydro-electric which can be turned on and off quickly to smooth out supply or meet peak demand; smart electric meters can be used in homes to automatically recharge electric vehicles when demand is low.

D.7.3.4 Timetable

With sensible policies, nations could initially set a goal of generating 25% of their new energy supply using WWS sources in 10 to 15 years. This goal would be gradually raised to the point where nearly 100% of new supply would be generated by WWS in 20 to 30 years. Under a very aggressive policy, all existing fossil fuel capacity might be retired and replaced in the same period; under a more modest and likely policy, full replacement might be completed in 40 to 50 years.

D.7.3.5 Fact or fiction?

Critics have responded that the study is riddled with errors and inconsistencies, and flagrant use of very optimistic assumptions. For example, with only 6 billion people, there may not be enough rooftop space in the entire world to house the *1.7 billion solar rooftop solar systems* called for in this study.

One critic calculated that approximately 15 square miles of land are needed to generate 1000 MW with solar thermal and voltaic plants. The 90,000 solar plants (rated at 300 MW apiece) would require some 450,000 square miles, nearly the size of Texas and California combined. Moreover, the solar mirrors and panels must be washed once a week, which would require massive quantities of maintenance and water. Furthermore, a wind turbine generating 5 MW would probably consume about the length of two football fields and would tower perhaps 80 stories; the 4 million windmills (rated at 5 MW a piece) required by the plan would extract a huge ecological footprint, hardly a benign source of energy.

What is interesting is that the WWS authors barely mentioned the use of nuclear power, the one widely proven technology that can produce green energy. One critic observed that the authors of this proposal barely make any distinction between nuclear and fossil fuels, which were lumped together as the old technologies. Nuclear energy's advantage is its energy yield per pound of resource consumed. Simply put, a pound of uranium contains 2000 times as much energy as a pound of coal. This translates into a 110-car train loaded with coal, arriving every 30 hours to fuel a plant, versus 6 tractor trailers loaded with uranium fuel rods arriving once every 18 months. Another criticism is that the WWS authors of this study claim that nuclear power emits 25 times more carbon over its life cycle as wind. But this assertion is debatable. A nuclear reactor contains about 500,000 cubic yards of concrete and 120 million pounds of steel. In contrast, a single 45-story wind turbine stands on a base of 500 cubic yards of concrete and contains as much metal as 120 automobiles. As 2000 of these are equal to one large nuclear reactor, this adds up to twice as much concrete and steel, and translates into significant quantities of greenhouse emissions produced over the construction phase (mining, concrete curing, forging, transportation, and construction).

Every form of electrical generation has a "capacity factor," the percentage of time, on average, the plant is up and running. All plants go periodically on- or offline (maintenance, refueling, or simple unavailability). Of *all* energy sources, nuclear energy has the highest capacity factor (greater than 90%); solar has the lowest (20%). While the WWS authors were factually correct in stating that windmills are only offline for maintenance 2% of the time, they neglected to mention that wind only blows about 30% of the time, resulting in a very low capacity factor.

Endnotes

1. Handmer, J.W. and Dovers, S.R., 1996, "A Typology of Resilience: Rethinking Instructions for Sustainable Development," 9 Industrial and Environmental Crisis Quarterly: 482.
2. United Nations World Commission on Environment and Development (Brundtland Commission), *Our Common Future*, p. 43, 1987 http://www.un-documents.net/wced-ocf.htm.
3. International Council for Local Environmental Initiatives (1994).
4. Definition adopted by the International Chamber of Commerce from the book published by the International Institute for Sustainable Development, Business Strategies for Sustainable Development: Leadership and Accountability for the 90s (1992).
5. Sustainable Consumption, http://www.gdrc.org/sustdev/concepts/22-s-consume.html, accessed January 23, 2010.
6. Sadler, B., 1995. Towards the Improved Effectiveness of Environmental Assessment. Executive Summary of Interim Report Prepared for IAIA'95. Durban, South Africa.
7. Carter, Neil, *The Politics of the Environment*, Chap. 8, Sustainable development and ecological modernization. 1st Ed 2001, 2nd Ed. 2007.
8. World Commission on Environment and Development,1987:49.

9. World Commission on Environment and Development, 1987:15.
10. Agenda 221, Principle 15.
11. World Commission on Environment and Development, 1987:9.
12. Richard MacLean, Competitive Environment, "I Say Green, You Say Sustainable", September 2009.
13. United Nations, Report of The United Nations Conference on Environment and Development, *Agenda 21*, Rio Conference 27 principles. (Rio de Janeiro, 3-14 June 1992) Annex I, Rio Declaration on Environment And Development, (Rio de Janeiro, 3-14 June 1992).
14. The National Environmental Policy Act of 1969, as amended (Pub. L. 91-190, 42 U.S.C. 43214347, January, 1, 1970, Sec. 2, "Purpose."
15. The National Environmental Policy Act of 1969, as amended (Pub. L. 91-190, 42 U.S.C. 43214347, January, 1, 1970, Title 1, Sec. 101(a).
16. E.O. 13514, Federal Leadership in Environmental, Energy, and Economic Performance, 74 FR 52117; October 8, 2009. http://www.archives.gov/federal-register/executive-orders/2009-obama.html.
17. Hargroves, K. and Smith, M. (eds.) (2005), *The Natural Advantage of Nations: Business Opportunities, Innovation and Governance in the 21st Century*, London: Earthscan/ James & James.
18. Marshall, Julian D., and Michael W. Toffel. Framing the elusive concept of sustainability: A sustainability hierarchy. *Environmental Science and Technology* 39, no. 3, 673–682 (2005).
19. Agassi, S., NY Times, December 2, 2009.
20. Elkington, J. "Towards the sustainable corporation: Win-win-win business strategies for sustainable development." *California Management Review* 36, no. 2: 90–100. (1994)
21. This section is based on information provided by The Global Footprint Network: http://www.footprintnetwork.org/en/index.php/GFN/.
22. This description was provided by the Global Research Development Center in Wupperthal, Germany and posted at: http://www.gdrc.org/sustdev/concepts/27-rucksacks.html.
23. Dow Jones, *Dow Jones Sustainability North America Index Guide Book*, Section 3.1 Version 2.0, August 2006.
24. The Worldwatch Institute, *2008 State of the World: Innovations for a Sustainable Economy*.
25. The Worldwatch Institute, *2008 State of the World: Innovations for a Sustainable Economy*, Chapter 13, "Investing in Sustainability,"
26. *Discover Magazine*, June 2010, pp. 47–51.
27. *Discover Magazine*, June 2010, p. 48.
28. *Discover Magazine*, June 2010, p. 48.
29. *Discover Magazine*, June 2010, p. 50.
30. Pickens Plan, http://www.pickensplan.com/theplan/, accessed January 24, 2010.
31. Pickens Plan, http://en.wikipedia.org/wiki/Pickens_plan, accessed January 24, 2010.
32. David R. Baker (2008-09-01). "Experts wary of Pickens' clean-energy plan." San Francisco Chronicle. http://www.sfgate.com/cgi-bin/article.cgi?f=/c/a/2008/09/01/MNO512K43O.DTL&type=printable. Retrieved 2008-10-04.

33 Natural Gas Vehicle Emissions." United States Department of Energy, Alternative Fuels and Alternative Vehicles Center. http://www.eere.energy.gov/afdc/vehicles/natural_gas_emissions.html.
34 "The Energy Report (Publication 96-1266). Chapter 11: Wind Power." Texas Comptroller of Public Accounts. 2008. http://www.window.state.tx.us/specialrpt/energy/renewable/wind.php.
35 Rob Bradley (2008-07-10). "Pickens' plan leaves U.S. energy security blowing in the wind." Institute for Energy Research. http://www.instituteforenergyresearch.org/2008/07/11/pickens-plan-leaves-us-energy-security-blowing-in-the-wind/.
36 Jacobson, M.Z. and Delucchi, M.A., A path to sustainable energy by 2030, *Scientific American*, pp. 58–65, November 2009.
37 Jacobson, M.Z. and Delucchi, M.A., A path to sustainable energy by 2030, *Scientific American*, pp. 58–65, November 2009.

Appendix E: Global climate change

E.1 The nature of the problem*

It began 4.5 billion years ago as a molten core, shrouded by a thick layer of hot gases. As the earth began to cool, its temperature eventually began to stabilize. But over the eons, the temperature of the earth's atmosphere would experience cycles of warming in response to both solar and earthly factors. Volcanic activity periodically released massive plumes of greenhouse gases that sent the Earth's temperature soaring. Likewise, collisions with killer asteroids sometimes turned day into night, resulting in freezing temperatures. Such events caused mass extinctions and altered the environment in ways that forced surviving creatures to adapt. The earth has experienced both long-term warming and cooling cycles, interspersed with episodic spikes of extreme climate change that have profoundly affected life. The National Oceanic and Atmospheric Administration's global warming website reports:[1]

> From the paleoclimate perspective, climate change is normal and part of the earth's natural variability related to interactions among the atmosphere, ocean, and land, as well as changes in the amount of solar radiation reaching the earth. The geologic record includes a plethora of evidence for large-scale climate changes. Massive terrestrial ice sheets throughout the Northern Hemisphere indicate cold conditions during the last glacial maximum (21,000 years ago). Warm climate vegetation, dinosaurs, and corals living at high latitudes during the mid-

---

* This chapter is based on Chapter 13 of the author's textbook, Global Environmental Policy: Concepts, Principles, and Practice (Eccleston & March) (2010).

Cretaceous (120–90 million years ago) indicate globally warm conditions. More recently during the Little Ice Age (roughly 1450–1890 CE), historic and instrumental records, predominantly around the North Atlantic, indicate colder than modern temperatures.

Perhaps it should come as no surprise that the earth may again be experiencing a change in climate. Whereas past changes have led to brutal extinctions, this change might also have equally profound implications. The major potential impacts of a modern global warming episode are gaining public attention: shrinking polar ice and glaciers, sea level increases that may threaten continental shorelines and island nations, and increasing average global temperature along with modified weather patterns in many areas. The causes and potential remedies to our global warming problem require at least an elementary understanding of the earth as a dynamic interactive physical-biological system—in effect a global ecosystem or biosphere.

The current debate is largely grounded on the premise that greenhouse gases (GHG), principally carbon dioxide (CO_2), are being pumped into the atmosphere. The preponderance of these GHG emissions is the result of burning fossil fuels such as oil, natural gas, and coal. As these GHGs accumulate in the atmosphere, they trap incoming solar radiation, warming the Earth's biosphere. Not surprisingly, there is a significant movement afoot to reduce the burning of fossil fuels. One of the principal drivers for adopting an ISO 50001 energy management system (EnMS) standard is that it provides an effective mechanism for reducing energy consumption and thus the generation of GHG emissions.

E.1.1 Global warming uncertainty and controversy

Perhaps no other environmental issue is as controversial as that of global climate change. There are advocates and opponents on both sides of the issue. Both sides can cite an armada of scientific evidence to support their position and claims. While the evidence is trending toward the position of anthropogenic (human-induced) global warming, this phenomenon is by no means proven or settled. Given the earth's variable history of climate change, it could also turn out that global warming may be the combined product of either human-induced and natural climatic variations, or other natural causes.

Much hangs on the scientific validity and verification of this theory. The cost of reducing greenhouse emissions to a level that will have a meaningful impact on global climate change will extract a very heavy economic burden on societies the world over; in hindsight, it will have been a grave error indeed, if efforts are taken to significantly reduce emissions, and it later turns out that greenhouse gases are having a negligible impact on our climate. Conversely, if greenhouse gases are profoundly affecting the Earth's biosphere, and nothing is done to mitigate these emissions, the results could be calamitous. Indeed, this is one of the defining issues of the 21st century.

E.1.1.1 The debate

The 2007 annual report by the International Panel on Climate Change concluded that anthropogenic climate change is *very likely* to produce a significant warming in global temperatures within our lifetimes;[2] here, the term "very likely" was defined as having a likelihood of occurrence of greater than 90%.[3] Many critics in the scientific community argue that the IPCC has overstated its case. Conversely, others such as James Hansen, a climate scientist from NASA, argue that climate change will be much more extreme than what the IPCC predicts.[4]

Emotions run high on both sides of this debate. Unfortunately, some advocates on both sides of the debate deny any and all evidence that runs contrary to their opinion. Such denials are not a phenomenon strictly limited to those who oppose the theory of global warming. Some scientists, who have shown a healthy skepticism toward global warming claim that they have been denied research funding, have been ostracized from the scientific community have been denied employment or promotions, and in some cases, have been terminated from their positions. This is hardly the way that the scientific method is supposed to work.

Prominent examples of dissenting literature include climatologist Patrick J Michaels' *Meltdown: The Predictable Distortion of Global Warming by Scientists, Politicians, and the Media,* published by the Cato Institute in 2005, and attorney Christopher C. Warner's *Red Hot Lies: How Global Warming Alarmists Use Threats, Fraud and Deception to Keep You Misinformed.* One reputable critic is Don Aitken, a political scientist and former Vice-Chancellor at Australia's University of Canberra.[5] Aitken contends that the proponents of climate change, including the IPCC, are a political creation rather than a genuine and objective scientific body. He asserts that the IPCC's scientific consensus appears to be manufactured, at least in part, by political pressure. Some dissenting IPCC scientists have made similar claims. Aitken also makes the reasonable point that the evidential basis underpinning the IPCC consensus is dangerously over-reliant on predictions generated by computer models; reality is much more complicated than our simple models. He goes on to argue that our current understanding of the global ecosystem fails to account for the many ways in which causal factors, relevant to creating the earth's climate, interact; thus, it is dangerous to presume that we accurately understand such matters.

Many critics assert, perhaps correctly, that attempting to interfere with the global climate process is problematic, and it is difficult to predict what the consequences will be. Global warming advocates counter that the IPCC computer models and scientific consensus are our best collective estimate of the potential effects of anthropogenic global warming; of course, their case isn't strengthened when former vice president Al Gore received a Nobel Prize for his documentary, *An Inconvenient Truth*, which has been conclusively shown to have nine noteworthy errors.

There are indeed many unanswered scientific questions and uncertainties that lead to different views of appropriate policy responses. Areas under debate, even by those advocating stringent policies to limit greenhouse gas emissions, include the:

- Degree, magnitude, and geographic distribution of potential climate change impacts
- Technical feasibility and affordability of the many technological innovations proposed as part of the policy
- Degree of potential adverse or beneficial economic impacts of proposed solutions
- Differing political, economic, and even religious ideologies among stakeholders

E.1.1.2 Framing the global warming policy debate

Spencer Weart, a physicist who specializes in science history, argues that when faced with scientists who publish global warming warnings, the public's natural response is to ask them for definitive proof and conclusions. When the scientists fail or are unable to respond with absolute assurance and conclusions, politicians habitually tell them to go back and do more research. In the case of global warming, waiting for absolute certainty could mean waiting forever; but the argument has been made that when society is faced with a new disease or an armed invasion, we do not put off decisions until more research is done. We act using the best evidence available.

Despite the fact that the evidence is moving in the direction of supporting anthropogenic climate change, there are indeed concerns regarding the efficacy of climate change policies. Hence, we can characterize the global warming policy problem as follows:

- What should policymakers believe concerning the reality of global warming?
- What course of policy action should be taken?:
  - Take no action because global warming is a hoax?

Appendix E: Global climate change 211

- Further monitor the data and improve our confidence in the nature and scope of global warming before committing to expensive remedies?
- Continue monitoring while taking policy action that we believe will ameliorate and ultimately reverse the global warming trend?

Regardless of the policy that is chosen, the problem of what scientists, statesmen, politicians, and the public accept as fact will not completely go away. Science is or at least should be an inherently open-ended process, scrupulous and candid in terms of exploring the evidence and performing experiments that validate or refute current hypotheses. Thus, there will always be room for dissenting views, especially by scientists and science-literate people. However, the problem runs far deeper than this. Powerful lobbyists and business interests, when threatened by proposed policy changes, often launch sophisticated public relations and media campaigns that can overwhelm the efforts of scientific experts. Conversely, the scientific community has been guilty of groupthink in the past, which has slowed the acceptance of more accurate theories, which run counter to established scientific dogma (Thomas Kuhn's theory, *The Structure of Scientific Revolutions*).

In the sections that follow, we will describe the issues and evidence for global warming. This does not mean that we endorse the current dogma. However, whatever their validity, understanding the theory of greenhouse warming is essential for those either dissenting or agreeing with the thesis of global warming. We will describe the scientific evidence supporting global warming, summarize the environmental impacts of no-action, outline the history of actions that have already taken place, review the technological opportunities for mitigating global warming and greenhouse gases, and describe the current efforts of international policymakers to deal with the problem.

One final note before beginning this examination: Control of emissions, effluents, and solid waste releases to the environment are already well embedded in the environmental management practices and infrastructure of most Western countries, albeit with varying degrees of effectiveness. Any policies aimed at controlling greenhouse gas emissions should therefore consider the existing pollution control measures that are already in place, and seek to leverage on them when feasible. To the extent possible, investments in greenhouse mitigation controls should seek to piggyback on proposals that also simultaneously produce economic benefits.

E.2 *Global warming causal factors and research*

The earth is a global ecosystem. All living things collectively constitute the living environment or biosphere. Animals feed on plants and other animal species. Plants feed on the remains of dead animals and plants. Some of the carbon dioxide emissions produced by human activities are absorbed by plant life in the oceans and on land. The remainder may be contributing adversely to climate change and other environmental impacts. Climate change stimulates additional environmental impacts that cause the global community to consider mitigation to reduce emissions, and adaption that reduces the effects on the socioeconomic system.

E.2.1 Gaia as a metaphor for the global ecosystem

James Lovelock proposed the *Gaia Hypothesis* as a metaphor for the earth as a holistic living self-regulating ecosystem. Lovelock argued that any creature adversely altering this balance triggers Gaia's self-regulating function. Lovelock's metaphor helps us understand that, in terms of a global ecosystem, the behavior of every living creature affects other creatures, all of which cumulatively impact the global environment. Hence, Gaia is an apt metaphor for the relationship between human civilization and the biosphere that it impacts.

E.2.2 The greenhouse gas factor

Greenhouse gases are a natural part of the Earth's atmosphere, but they are also being added by human activities. The principal source of human-generated greenhouse gases is that of burning fossil fuels, such as coal, oil, and natural gas. But other activities, such as deforestation and cement production, also play a major influence on increasing CO_2 levels. What proponents and critics can largely agree on is the increase in greenhouse gas, CO_2. As shown in Figure E.1, there is little debate that CO_2, the dominant greenhouse gas component, has increased significantly since the middle of the nineteenth century. Disagreements largely center on what this change in gas concentrations means and what effects it will have on future climate change.

Due to human activities, atmospheric CO_2 (as measured from ice cores) has increased over the past century from 300 to 386 parts per million (ppm), and the average earth temperature has increased approximately 0.7°C (or about 1.3°F).

A second point that proponents and critics can generally agree on is that there is a natural greenhouse effect that warms the earth. This greenhouse effect has existed for hundreds of millions of years and perhaps billions of years. Without this natural warming effect, the Earth would be in a deep freeze. Given what we know about the ability of greenhouse gases in warming the earth's surface, it is reasonable to expect that as

Appendix E: Global climate change

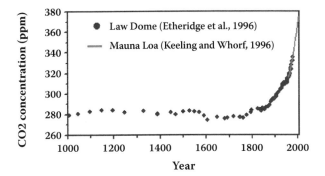

Figure E.1 Historic atmospheric CO_2 concentration. (From U.S. National Oceanographic and Atmospheric Administration.)

concentrations of greenhouse gases in the atmosphere rise above natural levels, the earth's surface will likewise warm. Many scientists have now concluded that global warming can be explained by a human-caused enhancement of the greenhouse effect. However, it is on this point that the debate generally ignites.

According to the IPCC Fourth Assessment Report, as of 2004, human activities are producing nearly 50 billion tons of greenhouse gas (GHG) emissions annually (measured in carbon dioxide equivalency).[6] Ambient concentrations of GHGs do not cause direct adverse health effects (such as respiratory or toxic effects), but public health risks and impacts as a result of elevated atmospheric concentrations of GHGs may occur via climate change.[7]

E.2.3 The greenhouse effect: How it warms the earth

The earth absorbs radiant energy from the sun and emits some of it back to space. The term *greenhouse effect* describes how CO_2, water vapor, and other atmospheric "greenhouse" (methane, etc.) gases affect the return of energy back to space, and in turn change the temperature at the Earth's surface (Figure E.2). Greenhouse gases absorb some of the energy reflected back from the surface, preventing it from being lost into space. The lower atmosphere is warmed in the process of absorbing this energy. The Nobel Prize winning chemist Svante Arrhenius first explained this principle in 1896.

Life on earth would be very different without the greenhouse effect. The greenhouse effect keeps the long-term annual average temperature of the Earth's surface approximately 32°C (or about 58°F) higher than it would be otherwise.

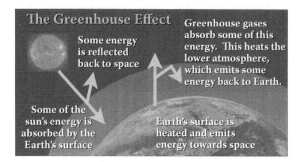

Figure E.2 The greenhouse effect. (From NOAA.)

E.2.4 Climate change research

As summarized later, there are several ways that climatologists study how the earth's temperature is changing. Instrumental temperature measurements, which extend the climate record back to the nineteenth century, provide a record indicating that the modern earth has warmed. These measurements indicate that the mean annual surface air temperatures have risen by approximately 0.5°C (0.9°F) since 1860. Over the last decade, however, this warming trend has leveled off and the Earth has actually cooled.

E.2.5 Paleoclimatology

The study of past climate is known as *paleoclimatology*. Instead of instrumental measurements of weather and climate, paleoclimatologists use natural environmental (or "proxy") records to infer past climate conditions. Paleoclimatic data generated from the study of past indicators such as tree rings, corals, fossils, sediment cores, pollens, ice cores, and cave stalactites are used to provide independent confirmation of the recent warming episode. The paleoclimatic record allows us to look at the variation of climate in the distant past.

Annual records of climate are preserved in tree rings, locked in the skeletons of tropical coral reefs, frozen in glaciers and ice caps, and buried in the sediments of lakes and oceans. These natural recorders of climate are called *proxy climate data*; that is, they substitute for thermometers, rain gauges, and other modern weather instruments.

The instrumental temperature record indicates that the earth has warmed by 0.5°C (0.9°F) from 1860 to the present. However, it is important to note that this record is not long enough to conclusively determine if this warming should be expected under a naturally varying climate, or if it is unusual and perhaps due to human activities.

E.2.6 Research based on proxy data

From a review of paleoclimatic data, both from thermometers and paleotemperature proxies, it is clear that the earth has warmed over the last 140 years. Paleoclimatic studies show that recent years are the warmest, on a global basis, over at least the last 1000 years. While each of the proxy temperature records differs, in part because of the diverse statistical methods, they generally indicate similar patterns of temperature variability over the last 500 to 2000 years.

Perhaps most striking is the fact that each record reveals a steep increase in the rate or spatial extent of warming since the mid-nineteenth to early twentieth centuries. They indicate that temperatures of the most recent decades are the warmest in the entire record. However, recent data suggest that this warming trend has leveled off and that the earth may have actually entered a cooling trend.

According to the IPCC, reconstruction of the paleoclimatic record provides relative confidence in the following conclusions:

- Significant warming has occurred since the nineteenth century.
- Recent warm temperatures recorded over recent years are among the warmest temperatures the Earth has witnessed in at least the last 1000 years, and possibly the last 2000 years.

E.2.6.1 Ice core evidence

Ancient ice traps bubbles of air, which contains CO_2 in the concentration that it was present in the atmosphere at the time the air bubbles were trapped in the ice. By studying ancient ice cores from Antarctica, it is possible to compare atmospheric concentrations of CO_2 that has been present in the atmosphere with temperature variations over the past 400,000 years (see Figure E.3).[8] A comparison of these two trends reveals a close correlation between the ways mean temperature has changed with respect to changing CO_2 concentrations. These fluctuations are nearly exactly mirrored across both the temperature and CO_2 plots over the last 400,000 years. Note that suddenly in the 1800s, as the Industrial Revolution began to take off, atmospheric CO_2 concentrations also begin an unprecedented upward climb, rising rapidly from 280 ppmv (parts per million by volume) in the early 1800s to the current level of 386 ppmv. This rise in CO_2 is greater than the highest concentrations previously attained in the course of the preceding 400,000 years. The post-Industrial Revolution mean temperature rose in tandem with the increase in CO_2 levels. Figure E.4 shows the correlation between rising CO_2 levels and mean temperature increase over the period since the post-Industrial Revolution period.[9]

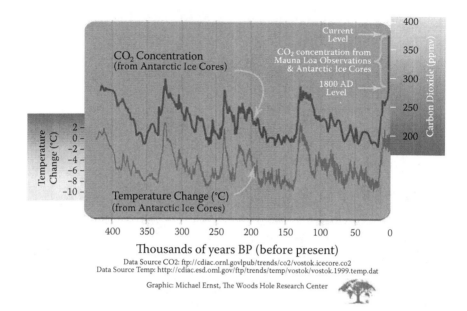

Figure E.3 Correlation between CO_2 and temperature. (From Barnola et al, 2003; Jouzel/Petit et al; Keeling et al; Neftel/Friedli et al. Data from Oak Ridge National Laboratory.)

Most experts do not contest the idea that some of the recent climate changes are likely the result of natural processes such as volcanic eruptions, Milankovitch Cycles (the earth orbital cycle), changes in solar luminosity, and variations generated by natural interactions between parts of the climate system (for example, oceans and the atmosphere). But what about human activities?

E.2.7 Some uncertainties

As discussed earlier, ice-core data appear to indicate that the increase in CO_2 is well beyond the range of natural variations. This increase correlates well with calculated fossil fuel emissions. Yet, there is uncertainty about some issues such as:

- What changes might future warming produce?
- Exactly how much warming has occurred due to anthropogenic increases in CO_2 versus other natural phenomena such as sunspot activity?
- How much warming can be expected in the future?

Arguably, the best available estimate is that perhaps 50% of the observed global warming is due to increases in greenhouse gas emissions.

Appendix E: Global climate change

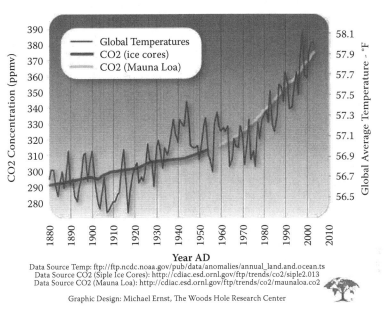

Figure E.4 Carbon dioxide concentration versus mean temperature measured by David Keeling and colleagues at Mauna Loa, Hawaii, and from polar ice cores, with average global surface temperature of the earth. (Image from Woods Hole Research Center.)

E.2.8 Alternative scientific explanation

The combination of solar variation, Milankovitch cycles, and volcanic effects is likely to have contributed to climate change, for example, during the Maunder Minimum. Apart from solar brightness variations, more subtle solar magnetic activity influences on climate from cosmic rays cannot be excluded, but neither can they be confirmed as physical models of such effects are still poorly developed.[10]

Solar activity measured by satellites and that estimated from the historical record using "proxy" evidence reveals that solar output varies (over the last three 11-year sunspot cycles) by approximately 0.1%.[11,12] Interpretations of proxy measures of variations differ. According to a US National Academy of Sciences study, variations in solar output have been shown to be the most likely cause of significant climate change prior to the Industrial Era. The relative significance of solar variability and other factors on climate change during the Industrial Era is an area of continuing research.[13]

A recent paper by Scafetta and West supports the thesis that solar variability is a major, if not dominant, factor pushing climate change.[14] They have shown an indirect connection between galactic cosmic ray

(GCR) intensity and global temperature.[15,16,17] This hypothesis suggests that the sun's changing sunspot activity permits more GCR to strike the earth during high periods of solar activity. The higher intensity of GCR may create more low-level clouds, which cool the earth. Conversely, when the solar activity is at a lower level, reduced GCR intensity may allow the earth to warm. Svensmark and Marsh have shown a close statistical correlation between sunspot activity and global cooling and warming over the past 1000 years. The recent rise in global temperature might be at least partially explained by the current lower level in solar activity.

A paper by Benestad and Schmidt contradicts the conclusions of Scafetta and West.[18] Many climatologists believe that the magnitude of recent solar variation is much smaller than the effect due to greenhouse gases.[19] Obviously, more research is needed to determine the precise effects that solar activity and other natural phenomena have on climate change.

E.2.9 Vital statistics on greenhouse gases

While CO_2 is the dominant GHG emission, on a per-molecule basis, methane and nitrous oxide are considerably more potent greenhouse gases. In fact, on a per-molecule basis, methane, as a greenhouse gas, is about 30 times more potent than CO_2. Some gases such as sulfur hexafluoride, a product of electrical insulation and magnesium smelting, are well over three orders of magnitude more potent than CO_2. Nevertheless, CO_2 is the dominant greenhouse driver due to its sheer volume. Figure E.5 shows GHG contributions from major economic sectors.[20]

A review of global emissions data indicates that China leads the world in total greenhouse emissions, while the United States leads in per capita emissions of CO_2. On a per capita basis, Canada closely trails the United

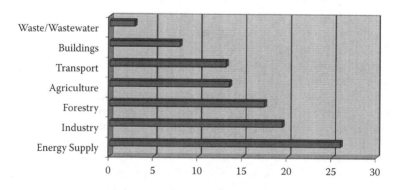

Figure E.5 GHG contributions from major sectors of the economy (%).

Appendix E: Global climate change

States, having the second highest emissions. A paper by Susan Salomon et al. concludes that climate change that takes place as a result of an increase in CO_2 is largely irreversible for a period of 1000 years after the emissions cease.[21]

E.2.10 Deforestation

Trees are natural consumers of CO_2. Destruction of trees not only removes these "carbon sinks," but the burning and decomposition of wood pumps voluminous amounts of CO_2 into the atmosphere. Some 20% of the world's GHG emissions are estimated to come from annual deforestation, such as in Brazil and Indonesia.

Many people know that China has now overtaken the United States as the world's number one emitter of CO_2. But few understand that because of its lack of deforestation policies, Indonesia is now the third largest source of CO_2, trailing only the United States and China.[22]

E.2.11 Methane

Based on existing climate models, it is believed that methane is responsible for about a third of the current warming trend.[23] Permafrost is filled with the remains of plant and animal matter that has been locked away in cold storage. Recent evidence suggests that the Arctic region is heating up twice as fast as the rest of the globe.

Thawing permafrost is like taking a package of frozen meat from the freezer and thawing it. As the meat warms, microbes oxidize it. On dry land, microbes convert this organic matter principally into CO_2. But in the wet, oxygen-starved depths of a lake, such as we find in the polar region, such thawing tends to release methane instead. As mentioned earlier, methane traps about 30 times as much heat as an equivalent amount of CO_2. Some climatologists are beginning to sound the alarm: melting ice and thawing permafrost might significantly increase methane release, which could significantly accelerate the rate of global warming.

E.3 Current and future impacts of global warming

Table E.1 presents some of the key findings and projections made by the IPCC.[24]

E.3.1 Findings from Worldwatch Institute

Relying mainly on IPCC reports and data, the Worldwatch Institute has made some impact projections of future climate change.[25] Warming

Table E.1 Some Key Findings and Projections Made by the IPCC, 2007

Observed changes in climate and their effects, and their causes
- Warming of the climate system is unequivocal.
- Most of the average global warming over the past 50 years is *very likely* due to anthropogenic GHG increases, and it is *likely* that there is a discernible human-induced warming averaged over each continent (except Antarctica).
- Global total annual anthropogenic GHG emissions (weighted by their 100-year GWP) have grown by 70% between 1970 and 2004.
- Anthropogenic warming over the last three decades has *likely* had a discernible influence at the global scale on observed changes in many physical and biological systems.

Drivers and projections of future climate changes and their impacts
- For the next two decades, a warming of about 0.2°C per decade is projected.
- Continued GHG emissions at or above current rates would cause further warming and induce many changes in the global climate system during the 21st century that would *very likely* be larger than those observed during the 20th century.
- Warming tends to reduce terrestrial ecosystem and ocean uptake of atmospheric CO_2, increasing the anthropogenic concentration that remains in the atmosphere.
- Anthropogenic warming and sea-level rise would continue for centuries even if GHG emissions were to be reduced sufficiently.
- Impacts are *very likely* to increase due to increased frequencies and intensities of some extreme weather events (heat waves, tropical cyclones, floods, and drought).

would be greatest in the high north and in the interiors of the continents. Warming would reduce snow cover, increase thawing of permafrost, and decrease the extent of sea ice. More frequent heat extremes and heat waves would be experienced, including more intense tropical cyclones, and heavier precipitation and flooding events in many regions. Precipitation can be expected to decrease in most subtropical regions but to increase in high latitudes. By 2050, it is projected that there will be less annual river runoff and water availability in some tropical areas.

Worldwatch cites several scientific studies in their assessment of the potential adverse impacts of global warming.[26] Coupled with a continuing increase in world populations, global warming would likely lead to increased human suffering and major loss of life in many areas. The world's socioeconomic systems would likely undergo radical changes as they adapt to the pace of the looming catastrophe.

E.3.2 US global change research program

The *US Global Change Research Program* was established in 1990 by the Global Change Research Act to coordinate US interagency federal research on climate change.[27] Thirteen US federal agencies and organizations participate in the program, with oversight by the White House Office of Science and Technology Policy, Office of Management and Budget, and Council on Environmental Quality. This program encompasses the US Climate Change Science Program, which synthesizes and provides up-to-date results on the science of climate change, including results from the Intergovernmental Panel on Climate Change.

The latest status report by the US Global Change Research Program, titled *Global Climate Change Impacts in the United States*, was issued to Congress in 2009. According to this report, the impacts of a changing climate are already being observed across the United States. The report provides a "state of knowledge" assessment of the science of climate change and climate-change-related impacts, now and in the future. It notes, "Observations show that warming of the climate is unequivocal" and "... is due primarily to human-induced emissions of heat-trapping gases."

E.3.3 The great climate flip-flop

Up until now we have concentrated on global warming. But our recent ancestors had different priorities, such as trying to start a fire to keep from freezing to death. While our hairy ancestors weren't busy hunting, or dodging wooly mammoths and saber-toothed tigers, they were braving something else: struggling to survive a frigid glaciation.

When one mentions the term *climate change*, it generally brings to mind the greenhouse effect with its killer heat waves. But paradoxically, as described below, global warming might paradoxically lead to a global deep-freeze, which could threaten the very fabric of our civilization.

E.3.4 Europe's furnace

The most populated areas of Europe actually lay about 10 to 15 degrees farther north than the most populous counterparts of North America. The sunny Mediterranean, known for its balmy weather, lies at nearly the same latitude as its chilly cousin Chicago. Europe's climate is abnormally warm compared to other areas of the same latitude in North America and Asia; its climate is moderated by warm water currents flowing north from the Tropics. This warm current keeps Europe about 5°C to 10°C (9 to 180°F) warmer in the winter than other countries at comparable latitudes.

If this warm oceanic current were to shut off, not only Europe but also much of the rest of the world would become much colder. Geological evidence indicates that the earth's temperature can change dramatically in just a few short decades or even less. One of the most striking examples occurred about 12,000 years ago as the earth was thawing from a long cold ice age. Something very strange happened. The gradual warming abruptly ceased and the earth crashed back into ice age conditions. But we're left with a puzzle: why?

E.3.5 The mighty conveyor belt

New evidence suggests that changes in oceanic circulation may well have been responsible for this rapid and dramatic turn of events.[28] A complex globally interconnected ocean current, collectively known as the *Oceanic Conveyor Belt*, governs much of our planet's climate. Depending on its temperature and salinity, a volume of seawater can float or sink into the ocean's depths. Water density varies with temperature, being greatest near its freezing point, making that volume of water more prone to sink into a warmer volume. As water warms up, it expands, decreasing its density, allowing it to float above a denser column. Salt content also increases density, making water more prone to sink.

Today, the oceanic Conveyor is driven by heavy cold-salty water from the North Atlantic Ocean. It is nothing short of a very long "salt conveyor." Warm surface water flows from southern to northern latitudes. Cold, dry winds blowing eastward from Canada evaporate the surface water as it approaches the northern latitudes. As this water evaporates, it leaves behind its salt load. As the remaining water becomes more concentrated in salt, it also becomes denser and begins to sink. As this water column sinks, it literally sucks more warm surface water from southerly latitudes to fill the void.

Like a gigantic conveyor belt, the North Atlantic also has a return loop running deep beneath the ocean surface. The circuit is completed as cold-dense water, having sunk, flows back toward the southern latitudes. Thus, a cycle or "conveyor belt" is formed. This process literally pushes water through the Atlantic Ocean.[29]

As far back as 1961, the oceanographer Henry Stommel was beginning to worry that the Conveyor system might stop flowing if too much fresh water were added to the surface of the northern seas.[30] If this happens, the Conveyor system could come to a screeching halt and would no longer carry heat from the tropics to the Polar regions. For example, more rain falling in the northern oceans could do this—exactly what has been predicted to result from global warming. Global warming might also melt Greenland's glaciers, flushing enough fresh water into the ocean surface to stop the Conveyor. Is this what happened 12,000 years ago as the continental ice sheets melted? Perhaps; the melting glaciers might have flushed fresh water into the North Atlantic, shutting down the Conveyor.

Appendix E: Global climate change

Between the years 1300 to 1850, the earth experienced a Little Ice Age. Some climatologists suspect that this frigid period resulted from natural climatic cycles that melted Arctic ice, causing the conveyor system to halt. Whatever the cause, the event caused massive suffering throughout Europe: famines, disease, crop failures, hypothermia, and perhaps even the rise of despotic leaders.[31,32]

While Henry Stommel's conveyor theory has evolved over time, the physics of adding fresh water to denser seawater is pretty clear. when denser water overlies less dense water, it sinks. In four decades of research, his theory has been improved but has not been seriously challenged. As Professor Wallace S. Broecker, one of the founders of this theory, observes: "The consequences could be devastating."[33] Agriculture on a global scale would be imperiled.[34]

E.4 Reducing greenhouse emissions

Figure E.6 shows a Worldwatch estimate of the potential energy from non-fossil-fuel technologies that can replace fossil fuels.[35] These technologies are all in commercial use today. Various efficiency measures and renewable energy technologies offer similarly large potentials for motored vehicles and the urban transportation sector. Because nuclear power produces no direct GHG emission, it might also play a role in reducing greenhouse emissions, even though its total energy and monetary costs are likely to be quite high. Green renewable energy systems, however may be ultimately cheaper and have greater potential to mitigate Greenhouse Gas (GHG) emissions.

With the advent of hybrid and all-electric vehicles, there is the prospect of a national system of recharge and battery exchange that could eventually reduce the need for fossil fuels in road transportation—assuming that renewable vehicle use increases. Agriculture also offers significant opportunities to reduce the methane and nitrous oxide components of GHG, and reducing deforestation can also significantly lower CO_2 emissions.

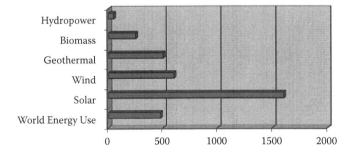

Figure E.6 Renewable energy technology potential (exajoules per year).

All this suggests that private investments and innovation, in some cases seeded by governmental incentives, can manufacture new energy devices that will eventually balance our GHG emissions with the earth's absorption capacity. However, as long as fossil fuel markets offer lower prices than alternative energy sources, renewables cannot compete, while the external costs of GHG emissions will soar.

E.4.1   The carbon capture and sequestration option

China's frantic construction of coal-fired power plants, now being built at the rate of one per week, could overwhelm any worldwide attempt to stabilize greenhouse emissions. While coal is much cheaper, its greenhouse emissions far exceed those of natural gas or even oil. China now uses more coal than the United States, Europe, and Japan combined, making it the world's largest emitter of greenhouse gases.[36] Given the large investment needed to convert to nongreenhouse emitting power sources, there is considerable interest in methods for removing CO_2 from coal or even the atmosphere itself.

E.4.1.1   Carbon capture

There are three principal ways to capture CO_2 from fossil fuel burning processed with today's technologies:[37]

- **Postcombustion Capture** involves a demonstrated method for treating flue gases (gas exiting the stack), but has yet to be demonstrated at a commercial power plant. A potential barrier is its high energy consumption.
- **Precombustion Capture** involves gasifying coal to remove CO_2 before it is burned in a coal-powered plant. This technology can capture between 80%–90% of CO_2. While it's far more efficient than postcombustion capture, it too has not yet been commercially demonstrated.
- **The Oxyfuel Process** is largely based on conventional power plant technology in the predemonstration phase. Combustion takes place in 95% pure oxygen, enabling efficient capture (99.5%) of CO_2.

E.4.1.2   Carbon sequestration

Carbon sequestration involves geological storage of captured CO_2 in ways that prevent it from being released into the atmosphere. After capturing the CO_2, it could be sequestered using one of the following methods:

- **Binding CO_2 to Silicates,** which is in the discussion stage, creates a stable solid that can be buried in the ground.

- **Fixing CO_2 with Algae to Produce Biomass** provides animal fodder, biodiesel, or construction material. This approach is at the discussion stage.
- **Pressurizing CO_2 to a Quasi-Liquid** and pumping into deep underground geological formations, deep nonexploitable coal seams, or aquifers. This technology is in the early demonstration phase.

Worldwatch concludes that the cost in energy to capture, transfer, and store CO_2, when added to the costs to acceptably reduce the whole range of environmental impacts of coal production will not permit large-scale deployment of carbon sequestering methods within the next 10–15 years. If this is the case, then China and other nations that continue to bring coal power plants online pose a serious challenge to any policies aimed at reducing global GHG emissions. Other engineering studies present a more optimistic assessment of this technology.

E.4.2 Plan G: The geoengineering option

Experience has shown that it is difficult to secure international agreements, such as the Kyoto Protocol, to limit GHG emissions. But Plan G, or the geoengineering option, has the potential to change the entire policymaking game. Instead of facing a dilemma where one large nation can foil the entire effort to curb global warming, a single nation has the potential to curb the effects of global warming single-handedly. For example, one option, such as pumping sulfur dioxide gas into the atmosphere, is a lot easier than trying to achieve consensus among 200 countries, each of which has strong economic incentives not to sign on to a protocol or even to cheat on their reported emissions.

Some policymakers view geoengineering as a last-ditch option. Others fear it could result in unpredictable and potentially calamitous consequences. Regardless of their views, most agree that geoengineering technology is probably powerful enough to reshape the planet's climate, and is so cheap and easily implemented that some may be tempted to take action into their own hands.

Some of the clues for a Plan G contingency come from nature itself, in the form of volcanic eruptions. Consider the 1991 eruption of Mount Pinatubo in the Philippines, which spewed sulfur dioxide and other pollutants into the upper atmosphere. These outgases were sufficient to absorb and reflect sunlight back into space, depressing global temperatures by about half a degree Celsius for several years. Many geoengineering options, such as those outlined later, have been proposed and more can be expected in the years to come.

E.4.2.1 Suspended fire hose method

Under this scheme, "fire hoses" can be suspended from helium balloons or zeppelins hovering 65,000 feet in the air to chemical factories on the ground. These factories would pump sulfur dioxide up through each hose at a rate of perhaps 10 kg/s. At the top of the zeppelin, the hoses would belch the sulfurous pollutant into the upper atmosphere. These sulfurous puffballs have the capacity to shield sunlight, cooling the planet as long as the process continues.

E.4.2.2 Whitening the clouds

Another Interesting proposal involves 'painting' the sky above the oceans white. A fleet of up to perhaps 1,500 ships would use large propellers to spray seawater high enough for the wind to loft it into the clouds. This spray would 'whiten' the clouds, turning them into reflectors that reflect sunlight back into space. By one estimate, the cost of constructing the first 300 ships would be $600 million, plus another $100 million per year to cover operational expenses.[38] This is but a small fraction of maintaining the U.S. Navy for a year.

E.4.2.3 Create forests of carbon-absorbing trees

The foregoing concepts fail to control the underlying and daunting problem of actually reducing CO_2 concentration. Every summer, plants absorb approximately one tenth of the atmosphere's CO_2 concentration. When these biological processes slow or cease in the winter, they release most of it back into the atmosphere. Some scientists propose using bioengineering techniques to create immense forests of carbon-absorbing trees to capture CO_2 and to keep it tied up in thick roots that would decay into soil, trapping the carbon.

E.4.2.4 Capturing carbon dioxide in the oceans

It has been observed for some time that natural plankton blooms can ingest large amounts of CO_2. One proposal builds on the observation that plankton blooms only occur when there is a sufficient supply of iron. This proposal involves seeding powdered iron across the ocean to stimulate massive blooms of plankton. But as with other geoengineering proposals, it is plagued with its own set of unknowns. For example, as the dead algae degrade, they might release methane. Methane is a much more potent greenhouse gas than CO_2.

E.4.2.5 Policymaking implications

Such proposals are well within the capability of a single rogue nation taking measures into their own hands. Imagine the plight of Bangladesh, whose population lives in coastal zones that could be flooded if the sea level rose. While Bangladesh is among the poorest of the poor, it could

Appendix E: Global climate change   227

easily take on such a program if it found its existence threatened. And who would have the moral authority to condemn them? Just as a ban on cloning in the United States shifted such research overseas, an outright international agreement to ban or severely restrict geoengineering might motivate some countries to implement potentially dangerous proposals on their own; such bans might simply constrain all but the least responsible nations.

Geoengineering solutions also provide governments with an easy excuse for not agreeing to limit their emissions. Or rather than cutting emissions today, policymakers can simply promise that in a future emergency they will support geoengineering measures to control emissions. Some policymakers argue that we should begin testing this technology now. There is always the possibility that at some time in the future we may have to consider such technology, and might need to implement it in a hurry. The least risky option would involve starting with small-scale field experiments and gradually ramping up the scale. With each technology option comes a host of scientific risks, as well as political, legal, and policymaking implications

E.4.2.5.1 Cost implications Geoengineering options are incredibly cheap. Compared to limiting carbon emissions, these proposals are a bargain basement sale, with some ranging from $1 billion to $100 billion per year. This compares to a cost of capping carbon emissions, which by some estimates might range a thousand times higher (on the order of $1 trillion annually).[39]

They also open a myriad of threatening possibilities such as the "Greenfinger" dilemma in which a rich "lone ranger" is as consumed with saving the planet as James Bond's nemesis Goldfinger was with gold. The world is now host to nearly 40 people worth $10 billion or more. Global geoengineering projects are within the financial wherewithal of a billionaire. Any billionaire could theoretically finance a program to reverse climate change single-handedly.

E.4.2.5.2 Risk factor This technology has the potential to unleash a Pandora's box of unknown ills. For example, schemes that depend on sulfur dioxide might produce acid rain, devastating large swaths of plant and fish habitat. Many concepts might radically shift climate and weather patterns. A change in weather patterns might benefit 4 billion people but devastate the agricultural system of 2 billion. Or perhaps a scheme cools the earth but unexpectedly and irreversibly destroys the fragile ozone layer.

E.4.2.5.3 Current status A 2009 study by a Rutgers University team, published in the journal *Geophysical Research Letters*, compared the pros and cons of the sulfur dioxide approach. The key pros included a cooler planet, and reduced or reversed melting of ice sheets and Arctic sea ice. The key cons included more droughts in Africa and Asia, continued oceanic acidification, and fewer blue skies. The team concluded that, given existing technology, the best method of lobbing aerosols into the atmosphere would be with the use of high-altitude military jets.

The view of the Institute of Physics is that geoengineering should "... be considered only as a last resort ... There should be no lessening of attempts to otherwise correct the harmful impacts of human economies on the earth's ecology and climate." The institute warned that geoengineering "should be seen as a prudent precautionary measure in case all other attempts to control dangerous climate change fail or are inadequate—for whatever reason." In 2009, the US House Science Committee held its first hearing on the implications of geoengineering.

E.5 Global climate policy

In this section, we review the history of how the global community has attempted to formulate effective policies. We also review policy ideas that are being proposed or are currently in debate. The challenges facing the many nations that are partners to this undertaking include:

- The sheer magnitude and complexity of the policy issues
- Significant scientific uncertainties
- The immense resources and investments required to take effective action
- The virtual certainty that the global human community cannot restore the preindustrial meteorological and ecological status quo in this century
- A host of economic, social, and political impediments to systematize and properly manage preventive and adaptive measures on a global scale.

E.5.1 Current status of policy: Cap and trade as lynchpin

The European Union published a less than enthusiastic assessment of its progress toward a fully functional climate change accord:[40]

> Topping the list of countries that have reduced greenhouse gases are **Germany (−2.3%)**, **Finland (−14.6%)**, and **the Netherlands (−2.9%)**. Bottom of the class in the Kyoto race are **Italy, Denmark,** and **Spain**. They are ranked bottom because they have increased

production in fossil fuel power plants. It appears as though these countries will not be able to meet their 2010 targets.

These figures do not stop the EU from observing this "progress" with optimism, based on new projections. The predicted GHG reduction for 2010 is far better than last year's predictions, and the EEA believes that even if current projections are maintained, the 15 EU member states will exceed the Kyoto Protocol's targets. It is true that 12 of the 15 member states expect to meet their initial objectives through a combination of national measures and European mechanisms. And there's the rub: only by putting in place and making use of additional measures, carbon sinks, and Kyoto mechanisms can this ambitious target be met; otherwise GHG levels will be reduced by only 4%.

E.5.2 The US Wadman-Markey cap-and-trade bill

The US Congress has seriously debated the most prominent version of a cap-and-trade bill, the *Wadman-Markey* bill. By any accounting measures, it will be an expensive proposition that could result in significant economic penalties in terms of such economic factors as unemployment and long-term growth. At the lower end, the US Environmental Protection Agency's own cost projection estimates this bill will cost the average American family an additional $1,100 per year. At the higher side, the Heritage Foundation estimates it would cost the typical American family from nearly $3,000 dollars annually and up to $4,600 by 2035.[41] The nonpartisan Congressional Budget Office reported that

> ... most of the cost of meeting a cap on [carbon dioxide] emissions would be borne by consumers, who would face persistently higher prices for products such as electricity and gasoline... [and] poorer households would bear a larger burden relative to their income than wealthier households would.

E.5.3 The economy and jobs

Critics respond that a rigid cap-and-trade bill would cost trillions of dollars and result in tens of millions of lost jobs. The Obama administration claims that millions of jobs can be created by expanding the U.S. electricity grid to accommodate new wind- and solar-power projects. However, a recent study by researchers at Spain's King Juan Carlos University questions whether "green jobs" are worth the public investments. Author Gabriel Calzada Alvarez cites Spain's investments as a case study. According to Alvarez, Spain has spent about $752,520 to create each "green" job, including subsidies for wind energy jobs. While jobs were created, other jobs were lost elsewhere in the economy. By Alvarez's estimate, the "U.S. should expect a job loss of at least 2.2 jobs on average, or about 9 jobs lost for every 4 created."[42] Of course, one also has to factor in the economic losses that could result from global warming if GHG levels are not stabilized.

E.5.3.1   Free rider problem

But more than the cost, many critics charge that so many allowances have been given away to special interests to gain support for its passage that the bill would be both expensive and only marginally effective. Unilaterally cutting American and European emissions amounts to giving away the sole bargaining lever for pushing other nations to also curb their own emissions. This is the *Free Rider* problem in its most blatant form. Is it realistic to believe that nations such China, India, or Indonesia will step up to the plate and sacrifice their own perceived economic self-interest simply because the U.S. did it first?

But problems with cap and trade do not end here. Consider the following—of what eventually could prove to be a long list—special favors to large lobbyists and powerful special interests.

E.5.3.2   Allocation problems

Perhaps no other power company uses as much coal as American Electric Power (AEP). AEP is one of the biggest electricity producers and users of coal in the United States. It seemed strange indeed that the chairman, Michael Morris, endorsed the Congressional Waxman-Markey bill, which would cap GHG emissions. The underlying reason for his support appears to have little to do with environmental health and just about everything to do with profits and self-interest. The bill requires many businesses to purchase GHG emission carbon credits. However, it would give away 85% of the credits initially. As a diverse company, AEP could conceivably tap three of the bill's free credits: 30% going to electricity distributors; 2% going to electric utilities; and 5% going to merchant coal generators.

If AEP accrues more credits than actual emissions, it can sell the extra credits, which translates into big profits. As a regulated utility (i.e., AEP faces no competition), the state government sets its rates. Thus, it can simply pass any additional costs onto consumers (it's a guaranteed win, no

Appendix E: Global climate change   231

risk proposition). Now for the entrée. AEP's interest in helping to craft the fine print of the Waxman-Markey bill explains the company's 11-fold increase in lobbying efforts and expenditures. AEP has been spending almost $1 million per month on lobbying for this bill. This illustrates how well-intentioned legislation can be reengineered by big corporate interests and lobbying. For every AEP that is exposed, one can only wonder how many others are incubating behind closed doors.[43]

E.5.4 Issues in global climate policy

Worldwatch cites the ethical issues that need to be dealt with in terms of controlling GHG emissions (see Table E.2).[44]

Worldwatch has compiled a list of proposals for adaptation, mitigation, and technology transfer likely to be considered by negotiators (see Table E.3).[45]

E.5.5 Concerns for cap-and-trade policy

The United States, while not yet a party to the Kyoto Protocol, has sought to introduce a cap-and-trade policy to reduce US carbon emissions. Lobbying efforts are mounting to gain competitive advantage over the potentially lucrative tradable emissions market that the US Congress is debating in 2009. Small companies are claiming that larger corporations are maneuvering to gain an unfair market advantage. Meanwhile, cement-makers, mining companies, chemical manufacturers, and food processors are angling for more carbon credits to help defer the costs of complying with cap-and-trade hardships. Smaller power cooperatives have already been awarded more permits by the US House in exchange

Table E.2 Ethical Issues that Need to be Dealt with in Terms of Controlling Greenhouse Emissions

- Can climate treaties be built on strong principles of fairness?
- How should rights to emit GHGs be allocated?
- Should the *Egalitarian Principle* be invoked, which states that every person worldwide should have the same emission allowance.
- Should the *Sovereignty Principle* be invoked, which argues that all nations should reduce their emissions by the same percentage amount.
- Should the *Polluter Pays Principle* be invoked, which asserts that climate-related economic burden be borne by the nations according to their contribution of GHGs over the years.
- Should the *Ability to Pay Principle* be invoked, which argues that the economic burden should be borne by nations according to their level of wealth.

Table E.3 Proposals for Adaptation, Mitigation, and Technology Transfer Likely to be Considered by Negotiators

India: Proposes a new global fund for adaptation. Industrial nations would contribute 0.3%–1% of their gross domestic product.

South Africa: Representing a coalition of African governments, proposes adoption of a program based on assessing costs from developed countries.

The European Union: Proposes expanding the global carbon market, leveraging private investment flows, and making financing predictable and tuned to the needs of developing countries.

Brazil: Proposes that developed countries finance a new Clean Development Fund that would finance the costs of climate adaptation for developing countries.

for a promise that they would not mount a grassroots effort against the cap-and-trade bill.[46] With such a lucrative market at stake, one can only imagine the infighting that would develop as such a market actually begins to unfold. This simply illustrates that environmental policy cannot be divorced from the day-to-day social, political, and economic factors that influence the policy arena.

E.5.6   China: The 800 pound gorilla in the global climate policy debate

Under President Obama, the United States may be moving in a direction of instilling "unilateral" reductions in fossil fuel use. Meanwhile, China sees fossil fuels as an essential ingredient for economic growth. At this moment, the Chinese are frantically scouring and purchasing rights to secure natural resources around the world. While China surpassed the United States in total CO_2 emissions, it currently shows no inclination to sacrifice economic growth for any emission limitation policy, as the United States appears to be doing. Between 1998 and 2008, China's oil consumption jumped 91%; its coal consumption 96%; its natural gas consumption 240%; and its overall energy use doubled. If anything, it appears that this pattern may only accelerate.[47]

E.5.6.1   China's one-child carbon offset credit policy

During the 2009 Copenhagen climate summit, Chinese delegates claimed credit for stifling some 400 million births since the 1979 implementation of their one-child policy. This policy generally limited couples to no more than one offspring, and was enforced by harsh government controls including coerced abortion and sterilization. By Beijing's calculations, its autocratic family planning program prevented 18 million tons of additional CO_2 from being released.[48]

Appendix E: Global climate change 233

Critics charged that it amounted to a trade-off: terminating lives to restrict carbon emissions. Peggy Liu, chairwoman of the Joint U.S.-China Collaboration on Clean Energy, defended China's argument for carbon offset credits; in a recent debate sponsored by the *Economist*, Liu argued that China must be given credit "for what it is not doing," explaining that "China's one-child policy reduces energy demand and is arguably the most effective way the country can mitigate climate change." Echoing this policy, Andrew Revkin, a *New York Times* environmental correspondent, even raised the prospect of baby-avoidance carbon credits. The Optimum Population Trust even announced a new campaign called PopOffsets which would pay poor Third World women not to reproduce; their underlying message is that people are the problem.[49]

Critics retort that this is simply the old specter of population control repackaged in terms of a new eco-faddism wrapped around the threat of global warming. Such critics point to the flamboyant predictions of overpopulation and doom voiced by prominent ecologists in the 1960s and 1970s that never materialized.

E.5.6.2 Climate change on the international stage

Climate change legislation under consideration in the United States, of and by itself, would have little effect on GHG emissions. Even a joint reduction by the United States and Europe may be insufficient to counter the significant increase in global GHG emissions. But what about negotiations at the international level?

Unfortunately, the prospects do not seem promising. Over 80% of the energy used worldwide comes from fossil fuels (petroleum, natural gas, and coal). CO_2 emissions are increasing at a rate of about 1% per year. Halting any further climate change "would require a worldwide 60%–80% cut in emissions, and it would still take decades for the atmospheric concentration of carbon dioxide to stabilize."[50]

Moreover, developing nations such as China, Brazil, and India are exempt from emission reduction requirements under the Kyoto Protocol, the main international agreement aimed at curbing global warming.[51] China is now the world's largest emitter of greenhouse gases. It currently builds new coal-fired power plants at the mind-numbing rate of more than one per week. Other "exempt" countries are also growing rapidly.[52] The efforts of the United States and other developed economies can have only a marginal effect on global warming if the developing world does not curb its emissions.

Then there is the "leakage" problem. Leakage refers to jobs that will migrate to China and other countries that have cheaper energy costs because they are not constrained by the need to limit GHG emissions. Because many industries will simply move their operations, there may be no net decrease in GHG emissions, even with stringent limits on emissions in developed countries.

E.5.6.3 Potential trade wars

Another problem that some economists point to is the historical parallels with the discredited Smoot Hawley tariff of 1930, which was designed to protect American industry and revive the economy from the effects of the Great Depression. Instead, this act spurred reprisals that led to an international trade war, aggravating the unemployment and widespread misery that ensued.

Steven Chu, President Obama's energy secretary, admitted in March of 2009 that "if other countries don't impose a cost on carbon, then we (U.S.) will be at a disadvantage."[53] To compensate for such a disadvantage, the argument is that the United States could impose greenhouse penalties (tariffs) on products purchased in other countries, notably China and India, that don't institute carbon emission controls. However, these countries would in all likelihood impose retaliatory tariffs on American exports that, like virtually all tariffs, would ultimately harm businesses, workers, and consumers in the United States.

In the end, the world could run the risk of sparking a 21st century global trade war that could severely damage the economies around the world. It is important to heed the lesson of the Smoot Hawley tariff of 1930. Rather than a unilateral carbon emissions ban by the United States and perhaps Europe, many critics counter that a more effective strategy would be to work on a cooperative international agreement by all principal players to a negotiated rollback in carbon emissions. However, countries such as China and India show no interest in reducing their greenhouse emissions. This circular reasoning leads one back to the potential for greenhouse penalties and tariffs.

E.5.6.4 The cap-and-trade policy dilemma

From a policy perspective, it has been argued that all nations must make reductions in GHG emissions. To this proposition, China and India have argued that the United States and other developed nations industrialized their economy with no limitations on greenhouse emissions; it is unfair to burden them with such restrictions before they have also had an opportunity to develop their own economies.

To this argument, the United States and other developed nations have countered that they largely developed their economies before evidence began to indicate the risk that GHG emissions pose; now that it is becoming increasingly evident that greenhouse emissions could profoundly affect global climate, such arguments by developing nations are no longer valid. Developing nations will simply have to industrialize using alternative energies. Meanwhile, many climatologists argue that if nations such as China and India continue along as they are, without making significant reductions, it could doom the planet.

In lieu of binding international agreements, some critics have suggested alternatives such as removing taxpayer subsidies for fuel use, eliminating regulatory barriers to nuclear power, and subsidizing biotechnology efforts to develop a clean fuel.

E.5.7 Uncertainty, the precautionary principle, and climate change

The *Precautionary Principle* is largely a political instrument rather than a strictly scientifically based policy strategy. This principle promotes a cessation or avoidance of activities involving uncertain and potentially high-risk human endeavors. In practice, it is often invoked when an influential interest group advocates an issue involving potentially risky and uncertain outcomes. If the interest group is successful in publicly raising concerns or fears, application of the standard scientific investigative method may be circumvented in favor of adopting immediate and stringent limitations of human actions. Government dictates, sometimes with little scientific basis, are imposed. Those who dissent from the popular and prevailing view may be vilified or even ostracized. Consider this in terms of Dr. James Hansen. Hansen, who directs NASA's Goddard Institute for Space Studies, is a leading authority and vocal advocate of global warming. He has stated "Global warming has reached a level such that we can ascribe with a high degree of confidence a cause-and-effect relationship between the greenhouse effect and the observed warming."[54] His scientific credentials are indisputable. He stated his scientific opinion on a well-grounded review of the scientific evidence. So far, so good. But here, particularly in terms of the Precautionary Principle, is where Hansen's views begin to cause some discomfort among his scientific peers and consternation among those whose views run contrary to his own.

Hansen criticizes the "natural skepticism and debates embedded in the scientific process," which he claims have been exploited by the fossil fuel industry as a means of downplaying the scientific world's confidence that CO_2 is fueling global warming. In *The Guardian*, he claimed that oil company executives have been active in spreading disinformation. Critics claim that his views have some elements of a conspiratorial nature, just as those on the opposing side of the issue have been accused of advocating conspiratorial views. In an article titled "Veteran climate scientist says 'lock up the oil men,'" Hanson was quoted as suggesting that those who promote the ideas of global warming skeptics should be "put on trial for high crimes against humanity." Statements like this only inflame the controversy surrounding global warming.

E.5.8 Skepticism and scientific scandal

In 1912, a paleontologist found bones in the Piltdown quarry. Known as the Piltdown Man, these bones, for 40 years, were heralded as the missing link between primates and humans. It took the scientific community four decades to realize that the Piltdown Man was nothing more than a hoax, and a bad one at that. Critics have long charged that much of the research on climate change was questionable, and that some scientists were guilty of scientific misconduct and unethical behavior.

The 2009 release of e-mails from East Anglia University's Climate Research Unit (CRU) has fueled allegations and controversy. These e-mails indicate a long and concerted history of scientific misconduct. Phil Jones, the CRU's director, e-mailed instructions to "hide the decline" in a recently observed cooling trend. For his part, Jones boasted that he had used a statistical "trick" to do just that.

Michael Mann, director of Penn State University's Earth System Science Center, and a principal author of the IPCC, developed the "hockey stick" graph purporting to prove that global temperatures were the warmest in recorded history. But in 2008, statistician Steve McIntrey exposed the flawed methodology. Other climate scientists were incredulous that Mann's work had passed two IPCC peer-reviewed rounds without anyone spotting these inaccuracies. Andrew Revkin, a senior *New York Times* reporter, was warned about Mann's discredited analysis; Mann replied in an e-mail to Revkin that "those ... who operate outside ... the system are not to be trusted." Instead of investigating the facts, Revkin supported Mann's position.

Other e-mails reveal efforts to ostracize skeptical researchers from the scientific community. Research that was inconveniently disagreeable to the CRC's mindset was ignored. Efforts were made to delete opposing research from IPCC reports. Some administrators had threatened to boycott journals that dared print research papers showing evidence that disputed global warming. Efforts were even made to withhold data and even delete some data that researchers did not want revealed to the public.

Most damaging of all is the fact that the CRU maintained a vital database of information, critical to the IPCC analyses and conclusions; the original data measurements had been "corrected." This is an accepted scientific practice, frequently performed to compensate for errors. However, the original uncorrected dataset was destroyed, purportedly because there was insufficient room to store it. At the very least, this is scientific sloppiness at its worst. But it implies something profoundly important to the science of climate change—there is now no means of verifying the accuracy of the original uncorrected data! This is particularly troubling as it calls into question data that is critical to the case of global warming, and upon which the world economies are gambling trillions of dollars over the coming decades.

Appendix E: Global climate change 237

In yet another embarrassing event, the IPCC stated in its latest report that "glaciers in the Himalayas are receding faster than in any other part of the world ... the likelihood of them disappearing by the year 2035 and perhaps sooner is very high." In 2009, four leading glaciologists prepared a letter for publication in the journal *Science* indicating that a complete melt by 2035 was physically impossible. In an interview, Professor Georg Kaser, who worked on the IPCC report, added fuel to the fire when he stated that he had warned that the 2035 figure was wrong in 2006, but his comments were ignored. The IPCC has since acknowledged that its report was in error in stating that the glaciers would melt away by 2035. In the *Daily Mail in London*, IPCC author Murari Lal stated that "we thought that if we can highlight it [melting Himalayan glaciers], it will impact policymakers ... and encourage them to take some concrete action." In calling for the resignation of the IPCC chairman, Andrew Weaver, a Canadian climatologist who coauthored three chapters in the IPCC report, stated "there's been a dangerous crossing of the line between neutral science and political advocacy."[55]

Errors such as these have led some commentators to claim that such events undermine the credibility of the IPCC if not the entire field of climate science research. Critics have dubbed this scientific coup, "Climategate." Sadly, such scientific misconduct, not to mention overt errors in the IPCC report, is the modern equivalent of scientific heresy. It has cast a shadow of suspicion over the entire discipline. How far this scandal will spread as of now is anyone's guess. It may take years for ethical scientists to reestablish the trust that has been lost among the public and policymakers.

E.5.9   Global policy implications

The issue of global warming is among the most complex and controversial scientific and environmental policy issues we face today. It is apparent that GHG emissions research needs to be funded and elevated to the highest priority so that uncertainties are reduced or confined to limits.

While attending the December 2009 global warming summit in Copenhagen, President Obama claimed that it had achieved an "unprecedented breakthrough." This was indeed strange given the fact that the Copenhagen summit committed no nation to any action, let alone a timetable for meeting such commitments. Leaders of developed countries were attempting to woo the leaders of undeveloped countries with promises of massive transfusions of cash, while at the same time considering how they would sell such explosive policies to the taxpayers back home.

Essentially, the Copenhagen agreement did virtually nothing. While some nations set their own goals on emission controls, stating that they would be internationally "codified," many critics contend that such "commitments" are next to worthless. Without something akin to universal cooperation, there is little hope of controlling GHG emissions—and there is virtually no hope of achieving universal cooperation.

So far, the public may not quite get what all this means: a recent poll by the Associated Press and Stanford University shows, for instance, that nearly twice as many Americans think action to contain warming would produce more jobs as those who understand it would reduce jobs.

Meanwhile, U.S. politicians and policymakers struggle with the anticipated fallout from taxpayer anger. Economic hardships are sure to follow any cap-and-trade bill passed by congress; or the U.S. Environmental Protection Agency's (EPA) attempts to enforce GHG Standness with the masses. Recently the EPA propagated a new rule requiring reporting of GHG emissions that exceed 25,000 metric tons or more of CO_2.

The seriousness of the occasion was occasionally punctuated by periodic moments of humorous folly, such as when the tyrant Hugo Chavez rose to address the summit. Chavez, who has crushed freedom and prosperity for the citizens of his own nation, Venezuela, stood up and denounced capitalism. As he did so, he undermined the largely free enterprise system (ostensibly about the global environment) of the developed nations who were being asked to foot most of the bill for curbing GHG gases. It will be difficult to garner international consensus among the world's taxpayers when despots such as Chavez command newspaper headlines, undermining the credibility of global summits.

Reaching an international consensus on reduction of GHG emissions increasingly appears to be a lofty but unattainable goal. Perhaps the ultimate policy solution is that of new technology, financed through a collaboration of partly private and government interests, and reinforced by tax and other economic incentives.

In the meantime, there is always the possibility that at some time in the future we may have to quickly consider the *geoengineering option*, the global equivalent of a silver bullet. If such an option is needed immediately, we may not have the luxury of performing carefully controlled experiments. If a silver bullet ever needs to be invoked, the least risky strategy may involve starting with small-scale field experiments now and gradually ramping the scale up to demonstration projects.

Endnotes

1 NOAA Paleo Perspective on Global Warming—Global Warming Sitemap: http://www.ncdc.noaa.gov/paleo/globalwarming/sitemapgw.html Last updated July 2009.
2 IPCC (2007). *Climate Change 207: The Physical Science Basis*.
3 IPCC (2007). *Climate Change 207: The Physical Science Basis*, pp. 22–23.
4 University of Oxford: http://www.practicalethicsnews.com/practical ethics/2008/04/policy-uncertai.html.

Appendix E: Global climate change 239

5 University of Oxford: http://www.practicalethicsnews.com/practical ethics/2008/04/policy-uncertai.html.
6 Synthesis Report at 38 (http://www.ipcc.ch/pdf/assessment-report/ar4/syr/ar4_syr.pdf.
7 74 Fed. Reg. at 66497-98.
8 Woods Hole Research Center, http://www.whrc.org/resources/online_publications/warming_earth/scientific_evidence.htm, October 16, 2009.
9 Woods Hole Research Center, http://oceanworld.tamu.edu/resources/oceanography-book/evidenceforwarming.htm, October 16, 2009.
10 UCAR (September 13, 2006). "Changes in solar brightness too weak to explain global warming." Press release. Retrieved 2007-04-18.
11 Weart, Spencer (2006). The Discovery of Global Warming. In Weart, Spencer. American Institute of Physics. Retrieved 2007-04-14.
12 Willson, R.C., and Hudson, H.S.,(1991). The Sun's luminosity over a complete solar cycle, *Nature*, 351, 42–44.
13 *Solar Influences on Global Change*, National Research Council, National Academy Press, Washington, D.C., p. 36, 1994.
14 Scafetta, N., and West, B. J., (2007). Phenomenological reconstructions of the solar signature in the Northern Hemisphere surface temperature records since 1600, *J. Geophys. Res.*, 112, D24S03, doi:10.1029/2007JD008437.
15 Svensmark, H. 2000. Cosmic rays and Earth's climate. *Space Science Reviews*. 93: 155–166.
16 Marsh, N., and H. Svensmark. (2003). GCR and ENSO trends in ISCCP-D2 low cloud properties. *Journal of Geophysical Research*. 108(D6): 1–11.
17 Marsh, N., and Svensmark, H., (2003). Solar influence on Earth's climate. *Space Science Reviews*. 107: 317–325.
18 Benestad,, R. E.; Schmidt, G.A. (July 21, 2009). "Solar trends and global warming." *Journal of Geophysical Research-Atmospheres*. doi:10.1029/2008JD011639. Retrieved 2009-09-07. "the most likely contribution from solar forcing a global warming is 7 ± 1% for the 20th century and is negligible for warming since 1980."
19 Houghton, J.T.; Ding, Y.; Griggs, D.J., et al., eds. (2001). "6.11 Total Solar Irradiance—Figure 6.6: Global, annual mean radiative forcing (1750 to present)," *Climate Change 2001: Working Group I: The Scientific Basis*. Intergovernmental Panel on Climate Change. Retrieved 2007-04-15.
20 Reformatted from Worldwatch 190. Original Source: IPCC.
21 Salomon, S. et. al. (February 10, 2009). Irreversible climate change due to carbon dioxide emissions. *PNAS*, 106(6): 1704–1709.
22 A study released Friday by the World Bank and the British government in 2007. "Indonesia is 3rd largest greenhouse gas producer due to deforestation," mongabay.com, http://news.mongabay.com/2007/0326-indonesia.html, March 26, 2007 (accessed October 16, 2009).
23 Simpson, S. (June 2009). The arctic thaw could make global warming worse. *Scientific American*, pp. 30–37. Special Editions.
24 Source: *Climate Change 2007: Synthesis Report* by the Intergovernmental Panel on Climate Change (IPCC). http://www.ipcc.ch/pdf/assessment-report/ar4/syr/ar4_syr.pdf.
25 Worldwatch Institute. *2009 State of the World: Into a Warming World*. Norton 2009.
26 Worldwatch 20-21.

27 DOE (September 1, 2009). *NEPA Lessons Learned*, Recent U.S. Climate Science Report—Useful Resource for Climate Change Impacts, pp. 12–13, Issue No. 60.
28 Rühlemann, C., Mulitza, S., Müller, P.J., Wefer, G., and Zahn, R. (1999). "Warming of the tropical Atlantic ocean and slowdown of thermohaline circulation during the last deglaciation," *Nature*, 402: 511.
29 Calvin, W. H. (January 1998). *Atlantic Monthly*, pp. 47–64.
30 Henry Stommel, Woods Hole Oceanographic Institution, in Massachusetts.
31 Lemley, B. (September 2002). The new ice age. *Discovery*, pp. 35–40.
32 Fagan, B. (2000). *The Little Ice Age*, Basic Books, New York, p. 246.
33 Newberry Professor of Earth and Environmental Sciences, Columbia University's Lamont-Doherty Earth Observatory.
34 Fagan, B. (2000). *The Little Ice Age*, Basic Books, New York.
35 Source: Worldwatch 132 (based on data from UNDP, Johansson et al., IEA.)
36 *New York Times*, May 11, 2009.
37 Abstracted from Worldwatch 99–102.
38 Wood G. (July/August 2009). *The Atlantic, Moving Heaven and Earth*, pp 70–76.
39 Wood, G. (July/August 2009). *The Atlantic, Moving Heaven and Earth*, pp. 70–76.
40 Source: European Union, *Kyoto Protocol—Results and Predictions* http://www.eudebate2009.eu/eng/article/28549/kyoto-protocol-predictions-facts-results-overview.html, accessed July 2009.
41 Heritage Foundation Center for Data Analysis (2009).
42 Burnham M. (April 16, 2009). Jobs at issue as labor-enviro coalitions stump for climate bill. *The New York Times*. http://www.nytimes.com/gwire/2009/04/16/16greenwire-jobs-at-issue-as-laborenviro-coalitions-stump-10548.html, accessed November 27, 2009.
43 Carney T. P. (September 23, 2009). How a power giant profits from greenhouse regs. *The Washington Examiner*, p. 15.
44 Abstracted from Worldwatch 161–172.
45 Abstracted from Worldwatch 182–183.
46 *Bloomberg News*, Greenhouse gas permits spark $100B lobbying fight, August 5, 2009.
47 *Daniel V. Kish is vice president for policy of the Institute for Energy Research.* September 20, 2009, http://www.washingtonexaminer.com/opinion/columns/OpEd-Contributor/U_S__-China-taking-widely-different-energy-paths-8264563.html.
48 Max Schulz (December 30, 2009). Population control: An ugly solution to climate change. *Washington Examiner*.
49 Max Schulz (December 30, 2009). Population control: An ugly solution to climate change. *Washington Examiner*.
50 Victor, D.G., Morgan, M.G., Apt, J., Steinbruner, J., and Ricke, K. (March/April 2009). The geoengineering option. *Foreign Affairs*, 88(2). Available online at http://www.foreignaffairs.com/print/64829.

51 Kyoto Protocol to the United Nations Framework Convention on Climate Change (1997). New York: United Nations.
52 Inman, M. (March 18, 2008). China CO_2 emissions growing faster than anticipated. *National Geographic News*. Available online at http://news.nationalgeographic.com/news/2008/03/080318-china-warming.html.
53 *The Washington Examiner*. Cap-and-trade would trigger a new global trade war, p. 2, August 6, 2009.
54 Fay J. (June 23, 2008). Veteran climate scientist says "lock up the oil men." NASA's Hansen goes DC, Posted in Environment.
55 The Examiner. EPA should put carbon regs on hold, p. 2, January 28, 2010.

Appendix F: Selected key ISO 50001 definitions

The ISO 50001 standard contains a list of definitions. This appendix defines some key terms that are particularly relevant in developing and implementing an energy management system (EnMS). The reader is directed to Chapter 3 of the ISO 50001 standard for a complete list of the definitions.

Energy For the purpose of the ISO 50001 standard, "energy" refers to the various forms of primary or secondary energy that can be purchased, treated, stored, or used in equipment, a process, or recovered for future use. With respect to an EnMS, it includes the fuels, electricity, heat, steam, compressed air, as well as renewables and other similar media.

Energy Consumption refers to the quantity of energy that is consumed.

Energy Efficiency is the ratio or other quantitative relationship between an input of energy and output of performance, service, goods, or energy. Examples include the conversion efficiency, energy required/energy used, or output/input.

Energy Performance refers to the measurable results related to energy use and energy consumption. With respect to the EnMS, results can be measured against the organization's energy policy, objectives, targets, and other energy performance requirements.

Energy Objective is the specified outcome or achievement established to meet the organization's energy policy in terms of improved energy performance.

Energy Target is the detailed and quantifiable energy performance requirement, applicable to all or part of an organization, that arises from the energy objective and that needs to be met in order to achieve this objective.

Organization This term refers to a public or private company, firm, corporation, enterprise, authority, or institution, whether incorporated or not, that has its own functions and administration and has authority to control its energy use and consumption.

Significant Energy Use refers to the energy use that accounts for substantial energy consumption or offers considerable potential for energy performance improvement. Specific significance criteria are determined by the subject organization.

Appendix G: The energy assessment

An energy assessment can identify steps that can be taken to significantly reduce energy consumption and lead to cost savings. Typically, the energy assessors will visit the facility building and interview facility managers and concerned staff, inspect lighting, air conditioning, heating and ventilation equipment, controls, air compressors, water-consuming equipment, and anything else that is using energy. The assessor will develop a list of energy conservation measures (ECMs) that have the potential to reduce energy usage and wasted money. Depending on the level of assessment effort, the assessors will quantitatively estimate how much savings potential there is for each of these measures, and the costs associated with implementation. Some measures will take decades to pay for themselves, while others will start paying for themselves almost immediately.

The results of the energy assessment can provide critical input necessary for preparing a comprehensive and rigorous plan for the ISO 50001 energy management system (EnMS). This chapter describes how the energy assessment is performed.

Typically, the larger the energy costs are in terms of a company's total operating cost (percentage), the greater the potential for cost savings. Facilities and companies that have performed only a cursory review of energy efficiency can often achieve 10% to 20% cost savings with minimal or no investment. Capital investments that involve paybacks within two years can frequently save an additional 20% to 30%.[1]

An energy assessment (audit) is a service that helps an owner accurately and reliably identify a building's energy, comfort, and health deficiencies. An energy assessment does for a large facility what the more familiar *energy audit* does for a private residence. The assessment evaluates where energy is excessively used, identifies changes that can be made, and prioritizes and recommends those changes to the company.

More than ever before, there is urgency to take actions and implement energy options that accelerate the shift toward sustainable development.

An energy analysis can broadly be divided into two phases: (1) collecting data and (2) evaluating data. Collecting data assists the assessors in quantifying energy flows into and within a facility. Evaluating data identifies potential energy efficiency opportunities; these opportunities can be prioritized based on cost-benefit.

Ultimately, the goal is to evaluate energy consumption and promote energy conservation. For example, the U.S. Department of Energy (DOE) administers a nationwide assessment program. The assessment is performed by a specialist employed by the DOE. This specialist will come to the facility location to perform an energy assessment. During this period, the assessment will be performed using specialized software developed specifically for the task. The process is entirely free to the company.

One of the chief prerequisites before the assessment begins is to make sure the company's management fully comprehends what is at stake. For instance, the DOE assessments described earlier typically leads to an average energy savings of $2.5 million USD per year. That is why it is so attractive. Frequently, many of the recommendations are relatively inexpensive to implement, but require a few extra duties, or a few minor inconveniences.

Case study

Recently, an energy audit was performed on a three-story, 85,000-square-foot office building built in the 1980s. The goal was to reduce the $25,000 monthly electrical bill. On completing an on-site evaluation of the building envelope, and electrical and mechanical systems, the assessment identified ways to save the owner $32,000 per year on the energy bill.

The assessment revealed that the structure was 95% out of compliance with modern energy standards. Electrical lighting inefficiencies amounted to $5,000 a month on the electric bill. Simply adjusting the lighting to comply with current code reduced the monthly lighting cost by $2,500.

The owner's budget did not allow for major retrofits, so the assessor concentrated efforts on the lighting systems and basic HVAC equipment.

The assessor suggests that additional savings are available with motion sensors, dimming features, daylight sensors, and LED existing lighting. If these features are installed in addition to the cost-saving measures already in place, the assessor estimated that it could reduce electrical costs to around $11,000 a month, resulting in a savings of more than half of their current cost.

Appendix G: The energy assessment

Example: LED saving solutions

LED bulb and tube retrofits have the following advantages: 80% less energy use than fluorescent tubes, extreme long life, dimming control, diverse light color selection, and mercury-free technology. It is important to note that not all LEDs are of the same quality, and an experienced assessor can objectively assist in making the best purchase for their particular needs.

Using modern LEDs, one can save 50% to 80% on lighting bills. Lighting generally accounts for 22% to 30% of commercial energy consumption, 37% for retail, and can run over 50% for certain facilities such as hospitals. In comparison, each office bay of fluorescent tubes uses perhaps $100 or more of electricity each year if they run from 6 AM to 6 PM. The latest generation of LED tubes only uses an estimated $20 of electricity for the same amount of lighting. For each of 100 lighting bays, this amounts to $8,000 in annual savings. The 80% energy savings can be dramatic. It is further enhanced by the fact that the bulbs last for 50,000 to 80,000 hours (over 25 years, assuming a typical use of 2,500 hours per year). In comparison, fluorescent bulbs may burn for 15,000 to 20,000 hours but they lose 60% of their efficiency after the first 12,000 hours. Commercial LEDs will literally run for decades, and the payback comes within two to three years, if not sooner depending on use.

G.1 The assessment process

Table G.1 outlines the typical process for performing a facility energy assessment.

Typical attributes that will be inspected and tested as part of the assessment include

- Lighting systems
- HVAC systems and controls
- Compressed air systems
- Renewable energy applications

Table G.1 Facility energy assessment steps

Conduct an initial screening to determine if an energy assessment will be effective
Form an Energy Assessment Committee to oversee the project
Determine the level of assessment to conduct
Perform initial data gathering
Develop evaluation criterion
Select an energy auditor
Conduct an internal awareness campaign
Perform the data collection and analysis
Report the review findings and recommendations
Implement changes

- Electric motors and drives
- Process systems
- Steam systems
- Heat recovery
- Building envelope upgrades
- Switching utility providers or utility rates

Table G.2 outlines the steps typically followed in performing a facility energy assessment.

Table G.3 outlines three general levels of assessments that are widely recognized. However, these three levels do not necessarily have clearly distinct boundaries.

In addition to calculating use and waste, a comprehensive energy assessment should also:

1. Estimate the fossil fuel energy resources that the facility uses on a yearly basis.
2. Estimate the carbon dioxide that is produced per year.

Table G.2 Facility energy assessment steps

Collect and analyze historical operational energy usage
Study the facility and its operational characteristics
Identify potential improvements and modifications that can potentially reduce energy usage and cost
Conduct an engineering and economic analysis of potential improvements and modifications
Prioritize a rank-ordered list of potential modifications
Prepare a report that documents the analysis process, results, and recommendations

Table G.3 The three general levels of assessment

Level I Assessment—Preliminary Energy Use Analysis and Walk-Through Analysis
This is a simple assessment based on available documents and information, physical inspection, and staff interviews. The assessment is used to create a baseline and identify obvious energy efficiency measures that can be easily implemented.
Level II Assessment—Energy Survey and Analysis
This assessment reviews existing and newly collected data to identify energy-saving measures and high potential measures for further investigation.
Level III Assessments—Detailed Analysis of Capital-Intensive Measures
This assessment is also referred to as an Investment Grade Audit. It is used to provide a detailed assessment of costs and benefits derived from capital-intensive energy conservation measures.

Appendix G: The energy assessment 249

Other energy-related impacts can, and often should, be evaluated. Understanding these impacts can not only lead to environmental improvements but also have the potential to save energy. For example, water use can indirectly involve significant cost and energy expenditures, yet it is often not included in an energy assessment.

G.1.1 Embodied energy

Embodied energy is the energy required to construct a given product from *raw material*. It accounts for the individual components of the product and the energy required for manufacturing. When the materials derived from nature are tracked through its entire life cycle (mining, manufacturing process, product development, and eventual recycle or disposal), the embodied energy is referred to as a *life cycle assessment*. Typically, the lower the embodied energy of a structure, the more environmentally friendly it is.

A house or office building tends to have a relatively large amount of embodied energy. All the parts of the building require raw materials from nature. The energy required to manufacture components such as the roof, windows, and floor adds to the total embodied energy. Transporting the individual components from the factory to the work site requires additional energy. Installing the parts of a house or building at the work site also adds to the total energy usage.

The maintenance that goes into the upkeep and operation of the building also adds to the embodied energy. Components that need to be replaced must be manufactured, using additional energy.

Once the structure has performed over its useful lifespan, the deconstruction and recycling of the materials occurs, adding to the embodied energy. Energy is required to demolish the building, load trucks with debris, and transport the material for disposal.

G.1.2 Energy management

Energy management is the process of monitoring and controlling the operating systems within a facility. The operating systems can include systems such as heating and air conditioning, ventilation, lighting, and power. While building, energy management techniques can be applied to a variety of different types of buildings, they tend to be most cost-effective when used in large commercial and industrial buildings. A building EnMS helps ensure that the building is operating at its maximum level of efficiency and performance. It is also important in regulating occupant comfort, and preventing poor air quality. These systems can be critical to the goal of minimizing wasted energy, which can significantly reduce expenses.

Most building EnMSs are operated using special software programs. These programs are capable of providing feedback on system operations and energy consumption. Many of these software systems allow operators to make changes to building automation systems. These energy control systems are usually operated by building management or maintenance personnel.

The software used to control building EnMSs can gather data from a variety of sources. Typically, they measure temperature changes, humidity levels, and occupancy patterns. This data is used to calculate energy use. Many systems also measure air quality to help maintain safe and healthy buildings.

G.2 Performing the energy assessment

The following steps are typically performed as part of an Energy Assessment. Figure G.1 illustrates a roadmap to energy savings, which should be integrated into the ISO 50001 EnMS.

Some recommendations for performing the energy assessment are depicted in Table G.4.

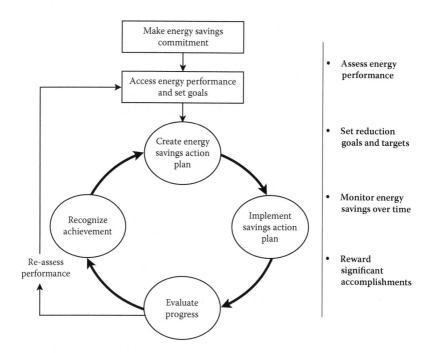

Figure G.1

Appendix G: The energy assessment 251

Table G.4 Recommendations for performing the energy assessment

---
Document each step of the process. Keep a notebook, as your notes can eventually become the energy-savings plan

Do not be shy about making suggestions to modify a process that is not part of the company's experience, culture, or goals

Display results using graphical formats

Start with a big picture plan and drill down into the details

Start planning capitalized projects (often, this requires seeking funding a year ahead of the budgeting process)

Recruit champions at the site level

Budget for time

G.2.1 Energy auditor's role

An energy auditor is much more than someone who compiles a list of obvious energy-saving tips. An energy assessor or auditor is someone who analyzes a facility's energy usage history and infrastructure. This may include doors, windows, roofs, insulation, the facility's energy-consuming equipment, HVAC, lighting, water systems, manufacturing processes, computer systems, and other energy loads to determine what energy efficiency measures could be implemented.

At present, there is no U.S. industry standard for the certification of an energy auditor. Several organizations provide certification and training. Among them are the U.S. Green Building Council, which established the LEED Professional Accreditation (LEED-AP), and the Association of Energy Engineers (AEE), which provides Certified Energy Auditor (CEA) and Certified Energy Manager (CEM) certifications. Additionally, associations and other organizational affiliations stressing the importance of energy efficiency include legal certification programs that award the licensed Professional Engineer (PE) and Registered Architects (RA) licenses.

G.2.2 Preliminary assessment and initial tour

The first step in performing an energy assessment should be a tour of the facility to identify what and how the assessment and measurements should focus on. During this initial tour, a thorough examination of the facility equipment and processes should be performed, and general observations should be noted. The assessment team identifies obvious problems and inefficiencies.

The team should identify any energy-related measurements that are routinely taken at the facility, for example, temperatures, fuel flow rates, and heating, ventilation, and air conditioning (HVAC) equipment.

If possible, determine if there is any evidence that the processes are not operating within expected ranges. Specific details regarding the use of any waste heat should be discussed with the facility operating personnel.

Following the walk-through, a follow-up with the equipment operators should be made to get more detailed information on how and when the equipment is run.

G.2.2.1 Involving staff

Experience has shown that energy-savings efforts will be most successful if your staff is involved in the process. Some important ways of involving the staff include the following:

- Appoint an internal energy manager responsible for managing and overseeing changes.
- Establish a team (with representatives from each company or facility department). Challenge the team with formulating and conducting an energy conservation program, which results in a predetermined level of energy savings.
- Establish team guidelines:
  - How they will assist in planning and participating in energy-saving surveys.
  - Develop record-keeping, reporting, and an energy accounting system.
  - Research, evaluate, and develop energy-saving ideas.
  - Establish challenges and rewards for reducing energy consumption.
- Communicate top-level management's commitment to energy conservation.

G.2.2.2 Analyze energy inputs

Data on gas, electric, and other utility bills is collected for the last one to two years. This data may be graphed to identify any trends. If gas and electric are the dominant energy sources, this may involve graphing monthly kilowatt and kilowatt-hour use for electricity and British thermal units (BTUs) for gas use. Other pertinent data such as monthly gas or electric costs may also be graphed.

G.2.2.3 Understand the utility bill

The facility's rate structure is key to understanding how energy is being billed. For example, for gas or liquid fuels, the rate is generally on a per-unit-volume (gallon or cubic feet) or thermal content (therm or BTUs in case of natural gas).

Appendix G: The energy assessment 253

Determining the rate structure of electricity is generally more complicated. For example, medium-sized facilities are often charged a rate based on a variety of factors. In addition to the simple energy charge for the amount of energy used, rates often include the demand charge (which varies seasonally), resource adjustment fees, power factor adjustment charge, conservation improvement program fees, and service fees.

Base load is non-weather-related use of energy in a facility. It includes such loads as lighting and equipment. It is important to isolate this variable to understand the effect that facility and process changes will have on energy use.

G.2.2.3 Gather data

Pertinent data is collected to help the assessor understand the operation and identify key focus areas. *Energy Benchmarking* provides a point of reference from which to make energy improvement comparisons.

Information on facility processes might be easily obtained from equipment vendors, and suppliers or organizations such as the Electric Power and Research Institute (EPRI) and the Gas Technology Institute (GTI). Moreover, the U.S. Department of Energy's Office of Industrial Technology (DOE OIT) web page provides an excellent source of energy efficiency information on many large industries and high-energy-consuming equipment.

Data on the geographic location, weather, facility layout, hours of operation, and the equipment list that affect energy use may also be needed. Understand the key safety issues of a facility before gathering information on the plant floor.

Data necessary for performing the assessment is collected. Such information may include data on HVAC, air compressors, motors, lighting, and specialized equipment and processes. This data helps assessors identify key energy efficiency opportunities. Examples are described later.

HVAC Review and inspect all HVAC equipment and collect data on unit specifics such as energy use, operating hours, and age.

Electric motors Review and inspect motors above a certain size (e.g., one horsepower) and prepare a data sheet on size and operating specifications. Such specifications may include full load amperage and average load, age, operating hours, and the average power factor.

Lighting Review and inspect lighting (including wattage and hours of operation). Consider tasks performed in various peak equipment loads. Identify infrequently used equipment or equipment that could be used during off-peak times.

Waste heat sources Review and inspect waste heat sources that have the potential to be used in a variety of processes such as supplying building heat during winter months or preheating water. Review all sources of waste heat: air conditioners, process cooling systems, boilers and heaters, ovens, air compressors, cookers, furnaces, etc. Attributes are measured such as flow rates and temperatures of waste heat sources.

Water heaters List fuel type and rate of energy use, daily demand, and operating temperatures.

Obtaining data might depend on using specialized instruments such as a combustion analyzer, light meter, ultrasonic air leak detector, and wattmeter.

G.2.2.4 Measurements

A measurement phase is obviously fundamental to performing an energy assessment. The importance of planning for this phase cannot be overemphasized. One important preparatory step often involves performing pretrip calibration and testing; this should be completed on equipment and sensors early enough to allow time for repair.

As discussed earlier, it is important to determine the data acquisition points that best support the mass and energy balances prior to making the measurement trip. In addition to location, timing is also an important consideration. Measurements should normally be made during normal operating conditions. It is essential that the data be collected at random points; this will help preclude the appearance of a trend that does not actually exist. Repeating measurements on subsequent days may also help to improve the accuracy of measurements during normal operating conditions.

After completing the measurements, it is important to complete a posttrip calibration of equipment and sensors to rule out the possibility of instrument drift that may have occurred during the measurement phase. This calibration step provides additional confidence that the data obtained is reliable and accurate.

Each project determines the specific hardware and software modeling tools that that will be used in the auditing process. Typical tools include

- kW meters, data loggers, BMS systems, and other tools to collect data.
- For energy modeling and financial modeling, a number of benchmarking/tracking tools exist that allow engineers to analyze the client's utility and operational expenses.

- For overall system performance and benchmarking, third-party software programs such as eQuest™ and Metrix™ may be used to model the building, calculate savings, and produce a baseline.
- Proprietary solution architect software can also help engineers and project managers collect, store, and manage audit data.

G.2.2.5 Example of a large office building

According to the Energy Information Administration, office buildings have more floor area (12.2 billion ft^2 [1.1 billion m^2]) than any other building type in the United States. They also have the highest total energy consumption (1.1 quadrillion Btu [1.2 EJ]). Moreover, the largest buildings have higher energy use intensity (energy consumption per square foot) than any other size of building.[2] While this creates large opportunities to save energy, it also creates challenges in terms of conducting energy audits in large office buildings. The auditor can be overwhelmed by the amount and complexity of the data. As a simple example, high-rise buildings frequently have unpredictable and uncontrolled airflows, driven by stack effects, exhaust fans, and higher-pressure air-distribution systems. The sheer size and complexity of such audits can lead to "audit exhaustion," resulting in a limited set of improvements.[3] While large office buildings present many technical challenges, they can also spur significant opportunities to save energy; the large building size can lead to economies of scale in terms of energy savings.

Beyond routine analysis of loads and equipment, the auditor should treat each building as unique. The auditor should look for anomalies in use, looking for ways to identify building-specific energy efficiency opportunities. This will generally include gathering actual HVAC operational data, equipment use, and flow rates, which can increase the understanding of building-specific problems and energy-savings opportunities.

Ambiguous audits or those lacking sufficient detail can lead to unclear recommendations. A general walk-through of a large office building will usually not provide sufficient data to identify significant energy savings. For example, in one instance, a walk-through audit of a large commercial office building identified savings of only 7%. A preliminary walk-through audit, in lieu of a comprehensive and in-depth audit, may cost the owner less but also runs the significant risk of delivering small savings, giving the false impression that large savings are not possible, and preventing the owner from considering a comprehensive audit, which holds the key to large-scale savings.

The final report should be a customized energy audit with specific recommendations and multiple improvement opportunities on a room-by-room or system-by-system basis. Reducing overlighting is a frequently missed improvement, so a light meter in the toolkit is essential. Lighting can offer great opportunities to save energy. It is not uncommon to find significant variations in lighting levels across various offices. Some offices may be significantly overlit, offering major opportunities for saving energy; the assessment also has other benefits such as revealing rooms that are underlit, which may adversely affect employee performance or pose health or safety risks. An education program can assist office occupants in using new technology such as double switch and switch-integrated photocell and occupancy sensors.

A "comprehensive energy audit" should include evaluating all energy loads and equipment within the building. This includes HVAC systems (in a commercial office building, typically chillers and boilers); HVAC distribution systems; lighting; plug loads such as appliances and computers; envelope improvements (walls, windows, roof, foundations, insulation); operation and maintenance improvements; and, of course, tenant education. On-site measurement of HVAC plant efficiency—including combustion testing for furnaces and boilers, and even testing of air conditioners and heat pumps—is becoming increasingly common. The energy audit should document recommendations in the audit report for implementing improvements.

So, how should a room-by-room audit be performed in a commercial office building? Generally, light levels and lighting inventories should be taken on a room-by-room basis; these measurements should also include noting occupancy levels and schedules for occupancy and lighting use. Room-specific HVAC attributes (distribution problems or mistaken temperature control setpoints) should also be noted. Information on plug loads (computers and office kitchen appliances) may be inventoried on a room-specific basis. Facility data sheets should be prepared to allow energy auditors, as they audit each room, to check off exactly what improvements will be evaluated in each room. Instead of providing general recommendations that are difficult to implement such as "Replace all lighting and install photosensors on fixtures close to windows," room-specific recommendations allow the work scope to be applied on a specific basis to the maintenance staff or the contractor.

Plug loads (computers and kitchenette refrigerators) can significantly contribute to electricity use. Their efficiency can frequently be increased; examples include setting display screens to turn off, or by implementing policies to turn off screens and computers. Plug loads require engaging tenants in energy efficiency.

The building stack effect and related infiltration losses can be reduced through weather stripping of windows and caulking window frames and other air-sealing methods. Such changes can also reduce the interior discomfort caused by airflow induced at the entrance to the building and on lower levels. Window replacements and storm windows can also reduce

Appendix G: The energy assessment

heat loss in the winter by half. Other improvements, such as interior or exterior shades, can further reduce loads.

G.2.3 Analysis and generating a list of energy efficiency opportunities

Evaluate energy efficiency opportunities for all operation areas in the plant. Compute the energy savings of proposed ideas and calculate a cost comparison. The Rutgers University Industrial Assessment Center provides benchmark figures that can be used in the energy analysis. New technology such as smart metering systems provide significant opportunities for saving energy, reducing energy impacts, and saving money (Figure G.2).

Audit results can be modeled as a baseline. Models can range from a simple spreadsheet for lighting to a whole facility energy model that estimates the savings that are possible from a full modification and retrofit. The savings are often compared to the preexisting facility baseline. It is sometimes useful to also compare these estimates to a new construction baseline because the new construction baseline is sometimes required by local building departments when a major retrofit occurs. Other resources for benchmarking results include government programs such as the US DOE/EPA's Energy Star Portfolio Manager.

Figure G.2

G.2.3.1 The simplified energy audit: assessing lighting and a simple HVAC system

A simplified energy audit may include

- Interviews with key staff to discuss occupancy schedules, and electrical consumption records, which may be reviewed at 15-minute intervals over a three-year period. Weather data may be collected at one-hour increments for the same time span.
- Light meter readings to identify each lighting type and reference to electrical consumption.
- Establishment of the electrical consumption baseline for the lighting system. Electrical consumption for pumps, motors, fans, etc., may also be reviewed.
- Installations of economizers and timers in the HVAC systems so that the air conditioning system can utilize free cooling whenever possible.
- Examination of energy codes to determine noncompliance levels.
- Development of a preventative maintenance program, to monitor filter changes, inspections, and pumps.

Figure G.3 shows a simplified computer estimate of alternative cost and payback for two purchased systems.

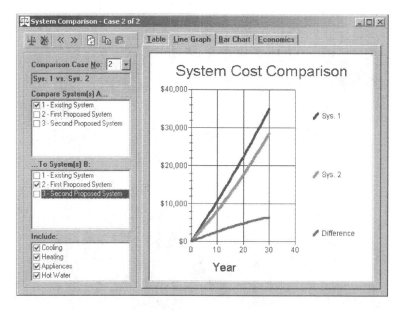

Figure G.3

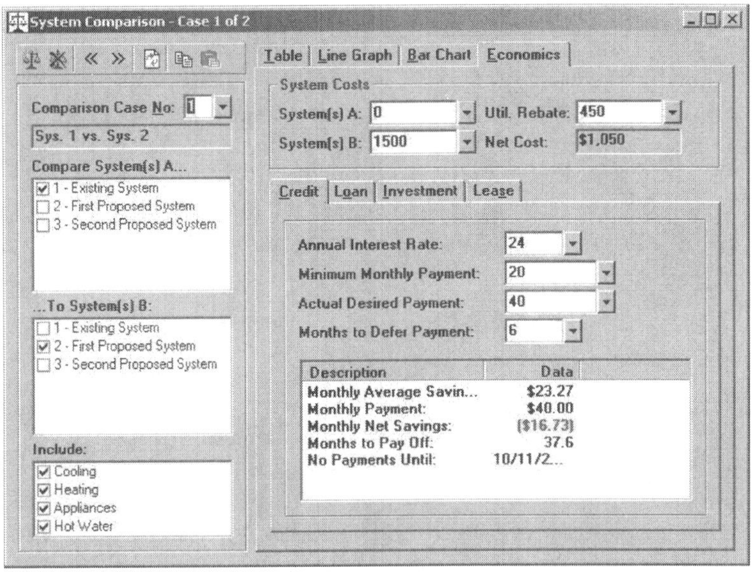

Appendix G: The energy assessment

Figure G.4

Figure G.4 shows a hypothetical computer estimate based on data acquired from the energy assessment, which compares the cost of the existing system with the payback period of Alternative 1 with Alternative 2.

G.2.3.2 Evaluate alternative ways to purchase systems

Consider alternative means of purchasing or retrofitting systems that can save your client money.

G.2.3.3 Preliminary mass and energy balances

After completing the initial facility inspection and reviewing the information obtained, mass and energy balance calculations are sometimes developed. The need for performing such calculations depends on the nature of the operational processes involved and the level and complexity of the assessment performed. This allows the team to identify the data required to complete a more detailed energy assessment audit. It can also highlight problems that may be encountered in collecting the necessary data. For example, there may be physical hindrances that prevent data collection from an ideal location; special arrangements may need to be made to ensure that the required data can be obtained.

Based on the preliminary energy balance, a list may be compiled that includes

1. Data and information necessary to complete a mass balance
2. Data and information necessary to complete an energy balance
3. Measurement instruments necessary to collect the data
4. Locations where measurements should be made
5. Identification of which measurements can or should be taken simultaneously
6. Estimated time required for each measurement

G.2.3.4 Identify potential energy-savings changes

Based on this analysis, a list of potential energy-saving changes is generated for review. These ideas are prioritized in the next section.

G.2.3.5 Tips for saving energy costs

- Adopt and implement more efficient industrial processes and equipment. Some studies have shown that this can result in a 24%–38% savings in energy use.[4] In manufacturing processes, the most promising approaches to improving efficiency are directed at motor drives, boilers and steam, lighting, and reducing air leaks in compressor lines.
- Make energy consumption a line item in the facility or company budget. This will encourage managers to make an effort to keep their costs within budgets and consider new ways to save energy and money. This can be an extremely effective measure when each department can measure its energy consumption.
- To the extent possible, recycle or reuse energy, particularly heat or heat as a by-product.
- Make energy savings one of the key factors considered in determining whether changes and improvements should be applied to processes, procedures, products, and production methods.

Some simple ideas for reducing energy use are depicted in Table G.5.

Table G.5 Simple ideas for reducing energy use

- Install and maintain caulk, weather-stripping, or other insulation on doors, windows, and air ducts to eliminate air leaks
- Turn off lights and other electrical equipment when practical
- Inspect and calibrate thermostats
- Where practical, install programmable heating/air-conditioning (HVAC) thermostats.
- Change or clean HVAC filters on a regular basis
- Recycle waste products

G.2.3.6 Cogeneration

Another energy-saving idea that is increasing in popularity involves *cogeneration*. In some circumstances where natural gas is readily available and affordable, **cogeneration** (Clean Heat & Power, or CHP) can be a very cost-effective energy solution delivering significant savings and reduced carbon output. A CHP can provide a particularly efficient way to generate primary or supplemental energy to run a manufacturing process. The CHP uses a fuel source, such as natural gas, to generate two forms of useful energy for the facility: electricity and thermal energy (either steam, or hot or cold water), which is a by-product of the electrical generation process. A CHP plant requires approximately 60% less fuel to produce the same amount of energy compared with a traditional plant utilizing a fossil mix/gas fuel source. Because CHP uses less fuel, as well as providing a cleaner source, emissions are reduced by as much as 53%.

G.2.4 Generate action plan and recommendations

Once the entire energy assessment has been completed, the team can recommend changes that may increase facility efficiency. Prioritized energy efficiency opportunities can be based on a cost-benefit assessment. If it is determined that the changes are cost-effective, the facility can implement the assessment team's recommendations. As necessary, bids are accepted and contractors are hired to implement the energy improvements.

A program should be developed to implement cost-effective opportunities and continually look for new opportunities. A program should be developed to communicate the energy-savings program with management and staff.

A follow-up assessment can quantify the effectiveness of the implemented changes. Normally, it is recommended to wait a period of time before making follow-up measurements to ensure the stability of the implemented changes.

G.2.5 Selecting a qualified assessor

A poorly performed commercial energy audit can cost more than it saves (even if the auditor offers it for "free"). It can be difficult to tell a good energy audit from a bad one. The problem is that assessors performing the audits are not required to possess any credentials. The person may have a four-year engineering degree and a great deal of experience in building construction. Conversely, the auditor may be an energy product salesman with no pertinent degrees or knowledge of building construction. The 7 key questions that should be considered in selecting an energy assessor are described in Table G.6.

Table G.6 Seven key questions that should be considered in selecting an energy assessor

---

1. What will the energy audit cost?

This is the single most important question. If the response is its "free" or "a few hundred dollars," you may want to skip the next six questions and look for a different assessor. No competent and well-qualified auditor is likely to give this work away.

Depending on the size and complexity of the facility, a professional audit should cost at least $1,000 (for a several hour walk-through and a verbal report of a single building). It may cost $20,000 or more for a several-day study of an industrial complex. Do your research, because these figures can vary widely depending on various factors and the complexity of the assessment.

2. What should the goal of my commercial energy audit be?

The goal should be reducing energy costs by at least 20% to 50%.

3. What will you examine?

The first priority should be identifying inexpensive opportunities for energy conservation.

4. What is your commercial energy audit qualification?

A commercial energy audit should be performed by a degreed energy engineer. He should have extensive construction knowledge and experience. It should not be done by a salesman.

6. How detailed will my energy audit be?

This is a difficult question. In reality, the assessor should typically focus on the 80-20 rule, that is, 20% of building conditions and energy systems and equipment that produce 80% or more of your energy bill. Their findings and recommendations should normally focus on that 20%, which ensures the best financial return on the energy investments.

7. How should the results be presented?

A commercial energy audit report should be a financial analysis. It should begin by discussing the opportunities identified to reduce energy costs. For each opportunity identified, it should estimate potential cost savings and benefits, specify the required products or equipment, and provide an estimate of the total costs. The recommendations should be prioritized based on each item's potential return on the investment.

Endnotes

1. Minnesota Technical Assistance Program, University of Minnesota. Accessed December 24, 2010, http://www.mntap.umn.edu/energy/assessment.htm.
2. Energy Information Administration, Office of Energy Markets 1. Forms EIA-871A, C, and E of the 2003 Commercial Buildings Energy Consumption Survey.

3. Shapiro, L. Energy audits in large commercial office buildings. *ASHRAE Journal*, January 2009.
4. Illinois Manufacturing Extension Center. Accessed December 23, 2010, http://www.imec.org/imec.nsf/All/Solutions_Source_Energy_Efficiency?OpenDocument.

Appendix H: Methods, tools, and techniques*

There are numerous tools, methods, and techniques that can be applied to implement an ISO 50001 energy management system standard. These tools include those available in the Quality and Value Methodology disciplines. From experience, there are a handful of tools, techniques, and methods that stand apart when considering implementation and management of a new standard or requirement. Organizations desire to avoid costly methods of implementing management systems that have low return on overall product or service value. Therefore, to justify their use, they need to show potential for improving the operation of the EnMS and demonstrate the ability to generate a return on investment. It is not uncommon to find that processes, system, programs, or procedures contribute little or no value to the bottom line. Thus, it is incumbent on an organization to formulate how compliance with standards and requirements will add value and ultimately help reduce negative impacts.

As in the case of "green" initiatives and sustainability, requirements have added cost to products such as buildings, homes, and consumer products. Many, if not most, customers in today's developed countries are concerned about the impact on the environment and are willing to pay additional costs knowing that in part these green products or services are helping reduce the carbon footprint and related environmental impacts. There are documented cases where such products have actually reduced cost or prices.

To understand what tools, techniques, and methods are the best for evaluating and implementing a standard such as an ISO 50001 EnMS, managers and professionals should become familiar with the variety of methods and techniques available in what is commonly referred to as process improvement tool kits. However, rather than mentioning

---

* This section was written by Mr. Bruce Lenzer who developed the Hybrid Function Analysis System Technique.

and explaining all of these available tools, techniques, and methods, the reader is directed to entities such as Goal QPC (www.goalqpc.com), which markets numerous pocket guides and memory joggers that explain various process improvement tools and methods. These resources are inexpensive and easy to read and comprehend. Outside professional services should be considered to help mentor and coach internal users on the use of such tools and aid in professional staff development. A few of the tools and techniques that can enhance the implementation of an EnMS are

1. Team Facilitation
2. Process Mapping and Process Management
3. Value Methodology [Value Management (VM)/Value Analysis (VA)/ Value Engineering (VE)]
4. Value Improving Practices (VIP)

Goal QPC markets pocket guides and memory joggers for the first three items listed earlier. The fourth method or tool set, VIP, has not yet been reduced to a pocket guide format. VIP consists of a set of decision analysis tools and techniques that encompass what is commonly referred to as a "Stage Gate" decision process. It also includes use of the first three tools techniques listed above. The VIP process allows for developing and validating the project scope in Stage One and completing with the final Stage Gate for approve implementation. During implementation, tools and techniques are used to monitor and resolve issues that arise during implementation. For large ISO 50001 projects, the VIP process or subset of tools may play an important role in implementing the strategy. However, as with any implementation, it is strongly advised to undertake a preliminary assessment or evaluation to determine which tools techniques, or methods are the best fit for a particular firm and their unique circumstances. For more information regarding the VIP Tool set, refer to the International Project Association (IPA).

Team facilitation is a foundational component when considering any type of standard or requirement. Using focus groups or multidiscipline teams adds great value for personnel to understand the impacts and opportunities for effective implementation based on their experience and knowledge of a particular organization. A skilled facilitator armed with a variety of tools, methods, and techniques can help teams produce high-performance results in a reasonable length of time.

Process Mapping lends itself to a variety of application tools. However, one contemporary innovative tool that stands out is *Hybrid Function Analysis System Technique* (Hybrid FAST). The Hybrid FAST takes the precedent function logic approach used in one of three styles of a FAST and may incorporate one or more of the styles. This technique was developed by

Appendix H: Methods, tools, and techniques 267

Bruce Lenzer and was awarded the Lawrence D. Miles Value Engineering Founders Award in 2010. Several companies and government entities such as Boeing, Flour Daniel, Parsons Engineering, AeroInfo Systems, and some Department of Energy Contractors have successfully trained staff in the use of Hybrid FAST as a tool used to map processes and improve them. Several projects were specifically scoped to address standards and requirements and ongoing value management of contract or regulatory compliance. Application of the Hybrid FAST tool begins with identification of the process, product, or service functions. Typically, functions are described in a *verb-noun* format. The verb is typically active, and the noun is something that is measurable. There are cases where a third word moderator, clarifier, or adjective can be used to better communicate the function statement. Functions are then oriented and described on a critical path, which in turn has subprocess activity, task, and decision functions that align in vertical format. This is conceptually much different from a typical flow diagram that usually reads input from left to output right. In the case of any FAST diagram or model, the input functions begin on the right and move to higher-ordered output functions on the left. The diagram reads "How" a function is performed from left to right. Asking the question "Why" when reading functions from right to left checks the precedent logic. If the question is not adequately answered by asking "How" and "Why" the function is performed, then there is a function logic mismatch that must be resolved. Logic-matched functions are oriented in the horizontal dimension of the diagram being constructed. In the Hybrid approach, the vertical axis includes activity and task functions that support the function in the horizontal axis being performed, and are oriented in a logical sequence and may have specific decision logic that must be addressed before moving to the next set of horizontal critical path functions. Functions on the critical path must also be performed to support the basic function of the product, service, or process. The basic function is the function that a product, process, or service must perform; otherwise there is no existence of a basic function for a product, process, or service. There are other interfacing techniques to a Hybrid FAST Diagram that enhance its value to a team or client. There are other companion tools that allow a facilitator and team to interrogate functions to determine time, cost, and organizational involvement in the performance of the function. Other tools, such as "Responsible, Accountable, Consulted, Informed (RACI)" matrices for developing a "Responsibility Assignment Matrix (RAM)," can be applied. Lenzer has developed an integrated tool that allows functions to be organized using their logic path on an x-axis and listing of organizational interfaces on the y-axis. At each interface, organizations can determine who has lead, support, or performing role responsibilities for the functions, and capture details about the organizational contribution to the process function. This helps when reviewing an "AS-IS" process or developing a new or "TO-BE" process. Such tools also aid in quantifying cost contributions to a process for cost modeling. Improvements can then be measured or simulated to determine the effectiveness or impacts of a variety of alternatives and process changes. In addition, the impacts include determination of overall value to the bottom line, and value in relationship to customer needs, wants, and expectations.

It is recommended to obtain consultation support when embarking on implementation of these tools and techniques. The consultation support should be a reputable entity that can provide a wide range of tools so that the right tools can be identified and applied to the appropriate scope or problem to be solved. The analogy is picking the right tool for the job (e.g., avoids using a sledgehammer to cut a board). Business entities typically expect to avoid circumstances where the wrong tools are used or where methods, tools, and techniques for such projects are ineffectively applied, resulting in poor-quality solutions or higher costs than expected. If applications result in such circumstances, then there is normally an impact to the expected return on investment.

For additional information on tools and techniques, the reader may want to obtain several Goal QPC Pocket Guides and Memory Joggers and obtain information available from SAVE International (formerly the Society of American Value Engineers), ASQ (formerly the American Society for Quality), and the Lawrence D. Miles Value Foundation. Resource websites are as follows:

1. www.value-eng.org
2. www.asq.org
3. www.valuefoundation.org

Appendix I: Applying the nominal group technique to maximize return on investment

Many organizations, particularly in manufacturing and industrial sectors, have very complicated energy generation/consumption systems or networks. One of the challenges in adopting an ISO 50001 Energy Management System (EnMS) is determining where emphasis should be placed so as to maximize return on investment. Interdisciplinary teams such as those involved in improving management systems represent a diversity of professional perspectives that may create contradictions. Experience has shown that this can lead to heated arguments over where scarce funds and other resources should be focused in improving management system performance.

The Nominal Group Technique (NGT) has proved to be useful in generating lists of ideas, and issues to focus on, sorting and then distilling them to a smaller manageable set for actual implementation. The NGT can help groups efficiently reach a consensus while ensuring that all opinions have been identified and given fair deliberation. NGT was pioneered by Delbecq and VandeVen,[1,2] and has been shown to enhance the effectiveness of decision-making in many different situations.

NGT differs from a simple traditional voting method where only the largest vote tally wins.[3] Group members silently write their ideas and issues as candidates for group discussion. For example, members may state their three top solutions for reducing energy wastage. These are pooled for a round-robin polling cycle. Thus, a large number of inputs can be discussed so as to foster open and comprehensive consideration.[4] Some solutions combine inputs from several persons and some are eliminated as the members vote to rank solutions. The number of NGT sessions required to reach consensus depends on the nature and complexity of the ideas, issues, and solutions and requires key participant presence at all sessions.

NGT groups frequently provide more unique ideas, more balanced participation between group members, and better generation of quality ideas and group efficiency.[5] The diverse and heterogeneous array of inputs can promote more effective decision-making.[6]

I.1 Typical NGT procedure

As shown in Table I.1, the NGT typically involves six stages.

Bartunek and Murnighan modified the NGT to better deal with ill-structured problems.[7] Under their modified process, ideas are generated

Table I.1 Six Distinct Stages of the Typical NGT Process

1. A qualified member of the staff (energy management representative) obtains the services of a facilitator, who will be in charge of facilitating the NGT process. Key individuals who will participate on the NGT panel are identified. The specific issues to be addressed and resolved are defined. Once the preliminary tasks have been completed, the Introduction step can be initiated.
2. *Introduction*: The group facilitator welcomes the members and explains NGT and the procedure that will be used. The facilitator is also responsible for ensuring that the process is as neutral as possible, avoiding judgment and criticism.
3. *Generation of ideas*: The facilitator describes the question to be addressed and requests each participant to independently write down ideas for resolving the question or problem. This stage may last as little as 10 minutes.
4. *Idea Sharing*: Each participant briefly explains his or her ideas. A cardinal rule during this phase is that all members must listen and that no member may criticize the ideas that are presented at this stage of the process. The facilitator records each idea on a flip chart. The round robin continues until all ideas have been presented. This process ensures that all group members are afforded an opportunity to make an equal contribution. This stage often takes as little as 15–30 minutes.
5. *Group discussion*: Participants are now invited to seek verbal explanation or further details about any of the ideas that colleagues have generated. The facilitator is responsible for ensuring that each member is allowed to contribute and that discussion of all ideas is thoroughly investigated without spending too long on any single idea. The group may suggest new items for discussion and combine items together or into categories, but no ideas can be eliminated at this stage. This stage typically can last as little as 30–45 minutes.
6. *Voting and ranking*: The final stage involves prioritizing the recorded ideas. The ideas are voted on. For instance, the voting may eliminate all but the top five ideas. These ideas are often ranked by priority.

and listed. The facilitator then questions if the ideas are relevant to the stated problem. If not, the problem is said to be *ill-structured*, and the generated ideas are combined into logical groups. These groups of ill-structured ideas are then treated as problems in their own right; the NGT procedure is then applied to each of them.

I.2   Advantages and disadvantages

One of the principal advantages of NGT is that it avoids two common problems that can occur during group interaction. Specifically, some members:

1. Tend to be reluctant to suggest ideas, even excellent ones, because they are concerned about speaking in a group or being criticized
2. Are reluctant to create or become engaged in group conflict

One of the most significant advantages of NGT, is that it tends to minimize both of these problems; examples abound where extremely useful solutions would have never come to the forefront had it not been for effective use of NGT. Another major advantage is that it frequently is a time-saving technique that allows a group to quickly identify and prioritize a large number of potential solutions, and then zero in on the most cost-effective ones.

One of the disadvantage is that NGT lacks flexibility because it is designed to focus in on one problem at a time. There must also be a certain degree of conformity on the part of team members. Occasionally, opinions may also fail to converge during the voting process.

Endnotes

1. Delbecq A. L., and VandeVen A. H, (1971). A group process model for problem identification and program planning. *Journal of Applied Behavioral Science* VII (July/August, 1971), 466–91.
2. Delbecq A. L., and VandeVen A. H, (1971). A group process model for problem identification and program planning. *Journal of Applied Behavioral Science* VII (July/August, 1971), 466–91.
3. Dunnette, M D., Campbell, J. D., and Jaastad, K. (1963). The effect of group participation no brainstorming effectiveness for two industrial samples. *Journal of Applied Psychology*, XLVII (February), 30–37.
4. Vedros, K. R. (1979). The Nominal Group Technique is a Participatory, Planning Method In Adult Education, Ph.D. dissertation, Florida State University, Tallahassee.
5. VandeVen, A. H., and Delbecq, A. L. (1974). The effectiveness of nominal, Delphi, and interacting group decision making processes. *The Academy of Management Journal*, Vol. 17, No. 4 (December 1974), pp. 605–621.

6. Gustafson, D. H., Shukla, R. K., Delbecq, A. L., and Walster, G. W. (1973). A comparative study of differences in subjective likelihood estimates made by individuals, interacting groups, Delphi groups, and nominal groups, *Organizational Behavior and Human Performance*, 9, 2, April, 280–291.
7. Bartunek, J. M., and Murnighan, J.K. (1984). The nominal group technique: Expanding the basic procedure and underlying assumptions, *Group and Organization Studies*, 9, 417–432.

Appendix J: The Fukushima Daiichi nuclear power plant disaster

The recent disaster of the Japanese Fukushima Daiichi power station (five nuclear reactors, three of which experienced partial me downs spewing radiation across much of the island) highlights the importance of rigorously investigating the impacts of any major energy proposal. A failure in EnMS planning and decision-making will almost certainly lead to poor results, if not dangerous decisions. The following case study illustrates how a misdirected planning process can lead to misdirected or failed decision-making.

How can future catastrophes be avoided? At the heart of the ISO 50001 energy management system (EnMS), lays the planning and decision-making process. An EnMS can offer an additional safety factor that can help compensate for flawed decisions or inherently poor management practices.

Performed properly, the EnMS planning process can lead to a significant reduction in energy consumption, cost savings for the organization, and improved environmental quality for the public.

Nuclear power license renewal

Most experts would agree that nuclear energy is potentially one of the most dangerous technologies in the world. The U.S. NRC's Division of License Renewal (DLR), directed by Brian Holian, is currently working full-speed ahead in re-licensing (more accurately referred to as "license renewal" or LR for short) an entire generation of aging reactors based on antiquated technological designs.

The existing fleet of U.S. nuclear reactors was originally licensed for a 40-year period. These licenses are expiring or nearing expiration.

The purpose of LR is to extend these licenses by an additional 20-year period, thus extending the operating window to 60 year period.

Given the priority and urgency at which the NRC is moving ahead with its re-licensing program, the entire fleet of antiquated nuclear reactors will be re-licensed within the next few years. What concerns critics is that the NRC performs a fast-paced and carefully choreographed process designed to show that the potential impacts of re-licensing an aging nuclear reactor are essentially benign. And this charge is reinforced by the following fact. Never—not even in a single case—has an aging reactor been denied a renewed operating license. As one NRC employee quipped, "No LR has been rejected, and I will be surprised if an application is ever rejected."[1]

The NRC's Division of License Renewal is responsible for issuing renewed licenses for the aging fleet of reactors. DLR's project branch (RPB1), under the direction of Bo Pham, is currently responsible for preparing the safety evaluation report and environmental impact statement (EIS) for re-licensing these aging nuclear power reactors. But, here lies the problem.

When an energy program has a troubling culture

DLR has suffered from morale problems and its own project managers (PMs) have complained of management and safety-related issues. The DLR morale problems became so significant that a decision was made to hold two focus group meetings to determine the root of these problems. An independent company was hired to facilitate these meetings. One was held to assess the opinions of DLR's project managers (PM) who are responsible for preparing the EIS (technically, a supplemental EIS) and safety evaluation report for re-licensing particular plants. The project managers' meeting was described as "somber." The DLR PMs were not shy about voicing critical and sometimes scathing comments. A summary of the comments includes the following:[1]

These are indeed serious allegations. But particularly disconcerting comments were those such as that managers are "bypassing the regulatory process and compromising the safety mission to impress upper management" and that DLR is "sacrificing quality for schedule." The results of the focus group meeting were not released to the DLR staff, let alone the public. Perhaps for good reason, Brian Holian never mentioned the results of the PM focus group to the DLR staff.

Perhaps more troubling, the chilling comments depicted in Table J.1 were not lodged by outside antinuclear critics, but by the very DLR PMs responsible for preparing the safety evaluations and EISs for reissuing the licenses. This deepens concerns. The managers are a vital part of the LR quality assurance process. For example, both Holian and Pham are

Appendix J: The Fukushima Daiichi nuclear power plant disaster   275

responsible for reviewing these EISs and safety evaluations and signing-off on their accuracy, rigor, and thoroughness. If DLR "… managers don't listen … act like know-it-all's," "are arrogant," are "schedule driven" with "dominant personalities," and make "poor management decisions," who is going to act as the critical stop-gap to ensure that a LR has been adequately investigated?

When energy schedules override public safety

As just witnessed, NRC's own PMs have charged that DLR is "sacrificing quality for schedules." The NRC has officially stated that an LR review be completed within an 18-month period. Schedules are now closely scrutinized and reported to upper management, to ensure that they meet the contrived 18-month schedule.

Now compare these management practices to the Japanese Fukushima nuclear disaster. Japan's experience clearly illustrates what can happen when schedules and sloppiness trump the quality of safety and environmental assessments.

NRC prepares an EIS for each license renewal it grants to an aging reactor plant. The EIS considers alternatives to license renewal; however, the public is often surprised to learn that none of these alternatives, not even the no-action alternative, have ever been given serious decision-making attention. Chapter 5 of the LR EIS (technically a supplemental EIS) investigates "severe accidents" of a nuclear accident. The EISs produced under the direction of the RPB1 management have routinely concluded that the risk from sabotage and beyond design-basis earthquakes (earthquakes the reactor is not designed to withstand) at existing nuclear power plants is "small." With respect to the assessment of a severe accident, Chapter 5 of the LR EIS have routinely stated that, "The probability weighted consequences of atmospheric releases, fallout onto open bodies of water, releases to groundwater, and societal and economic impacts from severe accidents are small for all plants." This is it. This is all the public and stakeholders get. Perplexed?

So how does DLR management justify the assignment of a "small" impact to a potential nuclear reactor accident? It employs a mathematical 'trick.' What RPB1 management cleverly does is to take the probability (which NRC argues is small), multiply it by the consequences (assume it is large), and then conclude that the human, environmental, and socio-economic impacts are small because the probability is so small. But let us step back for a moment and reframe the context of this problem. The EIS runs for hundreds of pages, examining every conceivable impact, from air emissions to water usage. Then, when it comes to a real issue. The issue that lies is at the heart of the licensing process. The issue of most concern to the public and supposedly at the center of the LR decision (i.e., the issue of safety and the potential consequences of an accident), it provides nothing but a cursory dismissal of the potential impact and a scant conclusion that the consequences of a large-scale accident would be "small"! Obviously DLR and RPB1 management is going to great lengths to avoid having to focus on the real issue that could kill the issuance of renewed nuclear operating license.

Let's reconsider the context of the DLR's decision-making process once more. Are the impacts of a potential accident that could result in direct radiation deaths, birth defects, evacuation of tens of thousands of down-winders, and near-permanent contamination of thousands of square miles "small"? Were the consequences of Chernobyl, Three Mile Island, and more recently the Japanese Fukushima Daiichi power station disaster "small"? If the probability of an accident is as small as the RPB1 manager appears to be claiming it is, why have there been four other near-catastrophic accident (near misses) in the United States, in addition to Three Mile Island? Again it is obvious that NRC management is going to great lengths, obscuring the facts, to avoid having to announce to the public and stakeholders that the impact of a nuclear accident could result in catastrophic human, environmental, and socioeconomic repercussions, as great or perhaps even greater than that experienced by Chernobyl or Japan.

A re-review of license renewal?

Given the management practices just described, what does such behavior say about DLR licenses that have already been issued? How many of these completed license renewals have suffered from a flawed process simply to meet licensing goals? Has RPB1 management already re-licensed a ticking time bomb? The answer is that nobody knows. A catastrophic meltdown may be next week or ten years from now. Some experts state that the impacts could be so severe that a major city or half a state could be 'lost.' It will require a full and comprehensive review of every renewed license issued to date to even remotely begin to answer this question. Every day that an aging rector runs with a perhaps flawed license renewal puts the public that much closer to a U.S. version of Chernobyl or perhaps Fukushima Daiichi. Imagine having to permanently evacuate the population of a major city and contaminating the better part of an entire state? Sooner or later the public we begin to understand how safety and accident issues have been twisted and obscured. It may be only a matter of time before Congress begins asking some very tough questions.

The goal of securing the environment and public health can only be guaranteed when an organization is truly committed to such a goal. A process designed to rubberstamp licenses on the shortest possible schedule possible will ultimately leave major question marks for the society

Appendix J: The Fukushima Daiichi nuclear power plant disaster 277

that must live with the consequences of such decisions. No management system, regardless of how well conceived, including ISO 50001, will make up for a façade, or an ill-conceived or slipshod energy licensing process. Nor can it completely compensate for poor management and decisions. However, when combined with a serious planning and assessment process it can at least help mitigate potential hazards that may result from ill-conceived or slack practices. An ISO 50001 can also be used to seriously investigate alternatives to unsafe technologies. It is to this end that this book has been written and dedicated.

In one license renewal project, two independent and separate projects were combined into a single project, for which a single safety evaluation and EIS were prepared. These projects (three plants in total) involved different types of plants, with different cooling systems, and different types of impacts; this complicated the process and substantially increased the amount of work required to complete the review. As manager of RPB1, Pham was warned by multiple sources that EIS could not be adequately completed within a contrived 18-month period. These warnings went unheeded. Consistent with some of the comments depicted in Table J.1, the RPB1 manager refused to listen and became agitated at being told that the EIS could not be completed on such a contrived schedule. As predicted, the project schedule began to slide. To reduce the slippage, this manager ordered the EIS contractor to stop further work. This was done despite the fact that the EIS contractor warned that the EIS was plagued with significant errors and inaccuracies. The EIS contractor appealed for more time to flush out errors and prepare a more thorough and accurate analysis. This feedback was disregarded. All of this was done, simply so that the EIS could be pushed out to the public to achieve the contrived schedule. In response to complaints that the EIS was plagued with errors, the RPB1 manager stated that "We'll fix it up later during the final EIS stage!"[1]

Table J.1 Summary of Key Statements made during the Project Manager Focus Group Meeting

- Some of the DLR managers are "very condescending."
- Managers are "bypassing the regulatory process and compromising the safety mission to impress upper management."
- "Poor management decisions" are being made.
- DLR "Managers don't listen—they act like know-it-alls."
- DLR "Managers are arrogant."
- There are "strained relations" between PMs and management.
- Managers are "schedule driven" and have "dominant personalities" (i.e., they place pressure on PMs to shortcut the process).
- DLR is "sacrificing quality for schedule."

[1] Personal Communications. Names withheld (2010).

But this is not the end of the story. The license renewal is a dual process consisting of issuance of (1) the EIS and (2) a safety evaluation to ensure that the plant can be safely operated for an additional 20-year period. At one point, the former safety project manager for this LR project (who asked to be relieved from the project in disgust) indicated that the safety evaluation was at least 4 and probably 6 months behind schedule, and continuing to slide even further behind schedule.[1] However, through managerial intimidation and shortcuts, the RPB1 manager was able to push the safety evaluation through the LR process and eliminate much of the delay. The safety review process was short circuited and failed to receive the sufficiently rigorous scrutiny that is normally expected to ensure that an aging plant can be operated safely for an additional 20 years. Again, this was done to meet a contrived project schedule in a vain attempt to impress upper management.

Evaluating the unevaluated

Under the U.S. National Environmental Policy Act (NEPA), an EIS is prepared to rigorously investigate the impacts of proposed actions and alternatives so that the public and decision-makers understand the full implications of taking an action that may significantly "affect the quality of the human environment." Chapter 5 of the LR EIS (technically a supplemental EIS) investigates "severe accidents" of a nuclear accident. The EISs produced under Pham's direction routinely conclude that the risk from sabotage and beyond design-basis earthquakes at existing nuclear power plants is "small" and that the risks from other external events are adequately addressed by a generic consideration of internally-initiated severe accidents. With respect to the assessment of a severe accident, Chapter 5 of the LR EIS routinely states that:

> The probability weighted consequences of atmospheric releases, fallout onto open bodies of water, releases to ground water, and societal and economic impacts from severe accidents are small for all plants.

This is all we get? Perplexed? Well, you are not alone. Pick up any LR EIS and turn over to Chapter 5 with its accompanying appendix (usually Appendix F) and read for yourself. So how does DLR management justify the assignment of a "small" impact to a potential nuclear reactor accident? It employs a mathematical 'trick.' What Pham's EISs cleverly do is to take the probability (which NRC argues is small), multiply it by the consequences (assume it is large), and then conclude that the human, environmental, and socioeconomic impacts are small because the probability is so small. But let us step back for a moment and reframe the context of this problem. The EIS runs for hundreds of pages, examining every conceivable environmental impact, from air emissions to water usage. Then, when it comes to a real issue that lies is at the heart of both the public and, supposedly, the LR decision (i.e., the issue of safety and the potential consequences of an accident), it provides nothing but a cursory dismissal of the potential impact and a scant conclusion that the consequences of a large-scale accident would be "small"!

Recall that the purpose of an EIS is to inform the decision-maker and public so that "informed decisions can be made." Are the impacts of a potential accident that could result in direct radiation deaths, birth defects, evacuation of tens of thousands of down-winders, and near-permanent contamination of thousands of square miles "small"? Were the consequences of Chernobyl, Three Mile Island, and more recently the Japanese Fukushima Daiichi power station disaster "small"? If the probability of an accident were so small, why have there been four other near-misses in the United States (in addition to Three Mile Island)? Perhaps more pointedly, if the impacts are really so small, why was the U.S. president and the nation in near-panic for days during the Three Mile Island accident? Was this because the potential impacts were "small"? Holian and Pham are clearly going to great lengths, obscuring the facts, to avoid having to announce to the world that the impact of a nuclear accident could result in enormous human, environmental, and socioeconomic repercussions.

A re-review of license renewal?

The DLR management is directly responsible for ensuring the completeness, accuracy, and thoroughness of the license renewal reviews. A management process that is "sacrificing quality and safety to meet schedules" and "poor management" is a dangerous formula, indeed. Given the practice just described, what does such behavior say about DLR licenses that have already been issued? How many of these completed LR projects may have suffered from a similar flawed process? Has RPB1 already re-licensed ticking time bombs? The answer is that nobody can be certain. It may require a full review of the renewed licenses issued to date to even remotely begin to answer this question. Every day that one of these aging reactors runs with a perhaps flawed license renewal puts us that much closer to a U.S. version of Chernobyl or perhaps Fukushima Daiichi. Imagine having to move the population of a major city and contaminating the better part of an entire state.

So what is the connection between the NRC's LR process and an ISO 50001 EnMS? Nations around the world have adopted environmental impact assessment processes, similar to that of the U.S. EIS procedure to evaluate the implications of taking a particular course of action. When the results of such a rigorous investigation or other similar planning and assessment processes are combined with an ISO 50001 EnMS, an organization has a powerful means of determining appropriate actions to take that can significantly reduce energy consumption while also minimizing the environmental footprint. But this goal can only be realized if the organization has performed an objective and rigorous assessment of the impacts and alternatives. A planning process that lacks rigor and objectivity cannot be expected to significantly contribute to the goal of reducing energy consumption and associated costs, or diminishing the environmental footprint.

Index

Number in () corresponds to ISO 50001 section number

A

AEP. see American Electric Power (AEP)
Agenda 21
    sustainability and energy policy, 184
    twenty-seven principles, 185
Allocation of resources, 125–126
Allocation problems, 230
Alternative energy, 144
American Electric Power (AEP), 230
Assessment
    levels, 248
    products' energy use over time, 93
Atmospheric carbon dioxide concentration, 213
Audit. *see also* Energy audit; Internal audit
    comprehensive energy, 256
    results from EnMS, 121

B

Base load, 253
Battery storage systems, 151
    advantages and disadvantages, 152
Biocapacity, 190
Biomass, 147
    energy, 148

C

Campbell, Colin, 168
Cap-and-trade policy, 228, 229, 231
    dilemma, 234
Capturing carbon dioxide in oceans, 224, 226
CAR. *see* Corrective action request (CAR)
Carbon dioxide
    atmospheric concentration, 213
    capturing, 224
    capturing from oceans, 226
    forests of carbon-absorbing trees, 226
    and temperature correlation, 216, 217
Carbon sequestration, 224
Checking performance, 97–116
    analysis (4.6.1), 97–98
    control of records (4.6.5), 114
    evaluating need for action (4.6.4b), 108–110
    evaluation of legal and compliance (4.6.2), 99
    implementing actions (4.6.4c), 111
    internal audit of EnMS (4.6.3), 100–103
    maintaining records (4.6.4d), 112
    measurement (4.6.1), 97–98
    monitoring (4.6.1), 97–98
    nonconformities and preventive action (4.6.4), 104–106
    records, 112, 114
    reviewing effectiveness (4.6.4e), 113
    reviewing nonconformities (4.6.4a), 107
China, 232
Climate change
    international stage, 233
    research, 214
Climate Research Unit (CRU), 236
Coal, 171
    fired generation, 139
Cogeneration, 261
Combination power plan, 196–198
Communications
    channels, 86
    commitment, awareness, and understanding of personnel, 84
    externally, 89

281

face-to-face dialogue, 87
implementation activities, 85
implementation and operation, 83–89
internally within organization, 83
package includes, 84
plan, 84
proactive personnel action, 87
responsibilities and authority, 25
Competence, 60
Comprehensive energy audit, 256
Comprehensive energy policy development, 173
Correction
    defines, 104
    example, 104
    steps for managing, 106
Corrective action request (CAR), 105
Cradle-to-grave approach, 191
Crisis preparing, 174
CRU. *see* Climate Research Unit (CRU)

D

Davis-Beese Nuclear Power Station, 274
Deforestation, 219
Democracy, 181
Design
    facilities, equipment, systems, and processes, 90–91
    input examples, 90
    requirements, 91
Disasters, 94
Division of License Renewal (DLR), 274
    managers and focus group meeting, 276
    re-review of license renewal, 278
Documentation requirements, 65
Documented record, 14
Documents
    control, 69–71
    legible, 75
Doe's peak oil assessment, xxvii
Dow Jones concept of corporate sustainability, 195

E

EA. *see* Environmental Assessment (EA)
Ecological footprint (EF), 190
Ecological rucksack, 191
    values, 192

Economy
    GHG contributions, 218
    and jobs, 229
EF. *see* Ecological footprint (EF)
Effectiveness of action, 98
EIS. *see* Environmental Impact Statement (EIS)
Embodied energy, 249
Energy, 243
    controversial conservation issues, 161
    documents, 68
    efficiency, 155, 243
    efficiency issues, 160
    embodied, 249
    objective, 13, 14, 15, 243
    objectives, 120
    source policy alternatives, 170
    target, 243
Energy assessment, 245–261
    assessment process, 247–249
    case study, 246
    energy audit assessing lighting and HVAC system, 258
    example of office building, 255
    gather data, 253
    identify potential energy-savings changes, 260
    list of energy efficiency opportunities, 257
    performing energy assessment, 250–261
    preliminary mass and energy balances, 259
    recommendations for performing, 251
    selection, 262
Energy audit, 245
    auditor's role, 251
    comprehensive, 256
    conduct, 102
Energy baseline
    energy planning, 48–50
    meaning, 49
Energy benchmarking, 253
Energy conservation
    consumption, 129–133
    and energy efficiency issues, 160
    generation, 129–133
    indicates, 155
    perspectives, 156–157
    sustainability, 129–133
    systems, 129–133
Energy consumption, 129–154, 243
    conservation, 129–133

Index

generation, sustainability, and energy
systems, 129–154
levels, 45
and production, 134–137
relationship to GDP, 134
section, 133
survey of sources, 139–152
sustainability, 138
Energy management system (EnMS), 249
audit results, 121
awareness, 61
communicated and made visible, 5
communicating importance, 12
consists, 1
criteria and methods, 26–28
Document Control System, 71
establish, implement, maintain,
and improve, 20
general requirements, 1–4
implementation, 124
versus improving energy performance,
xxiii
influence decisions, xvi
making comments and suggesting
improvements, 88
representative, 19, 21
requires investment of employees, 10
scope and boundaries, 10–11
tools and techniques enhancing
implementation, 266
Energy management team formation, 23
Energy performance, 243
awareness, 86
benefits of improvement, 63
changes within organization, 123
improvement implementation, 98
improvement opportunities, 90
improving and EnMS system, xxiii
indicators, 15, 51–52, 97, 120
objectives and targets, 13–14
quantify and measure, 15
refers, xxiii
Energy planning, 37–58
action plans, 56
activity review, 39
application and integration, 55
consideration of all factors, 56
consistent targets, 54
energy baseline (4.4.4), 48–50
energy performance indicators (4.4.5),
51–52
energy review (4.4.3), 42–46
general (4.4.1), 37–38

legal and requirements (4.4.2), 40–41
measures and time frames, 54
objectives, targets, and action plans
(4.4.6), 53–58
objectives and targets, 53
preengineering decisions for
baseline, 47
scheduled documenting and
updating, 57
selecting technologies and
methods, 47
Energy policy (4.3), 29–36. *see also*
Sustainability and energy policy
appropriate to energy use, 30
awareness, 61
changes due to management
review, 124
commitment to improvement, 29, 31
communication of policy, 34
complying with requirements, 32
comprehensive, 173
criteria, 30–36
documenting, 67, 68
establish, implement, and
maintain, 7
examples of sustainability, 196–202
framework, 33
information and resources, 32
management review, 120
support, 33
updating, 35
Energy production, 134–137
source, 135
Energy program manager management
representative, 25
Energy review, 44
analyzing, identifying, prioritizing,
and recording, 44
energy planning, 42–46
outputs, 97
recording and maintaining, 42
updating, 48
Energy use
effectively operating and
maintaining, 81
and efficiency worldwide, 158–159
ideas for reducing, 260
EnMS. *see* Energy management
system (EnMS)
Environmental Assessment (EA), 42
Environmental Impact Statement
(EIS), 42
Equity, 181

284   Index

Europe's furnace, 221
External documents
    controlling, 76
    defined, 76

F

Face-to-face dialogue, 87
Facility energy assessment steps, 247, 248
Failure modes and effects analysis (FMEA), 108
Fission nuclear energy, 144
FMEA. *see* Failure modes and effects analysis (FMEA)
Forests of carbon-absorbing trees, 226
Fossil fuels, 139
    advantages and disadvantages, 140
    comparison of energy sources, 160
Free rider problem, 230
Fuel cells, 153
Fukushima Daiichi nuclear power plant disaster, 273–278
    evaluating unevaluated, 277
    license renewal, 274
    NRC's troubled license review process, 274
    re-review of license renewal, 278
    schedules taking precedence over quality, 274–276
Fusion, 143

G

Gaia, 212
Galactic cosmic ray (GCR) intensity, 217–218
GCR. *see* Galactic cosmic ray (GCR) intensity
GDP. *see* Gross Domestic Product (GDP)
Geoengineering option, 225
Geopolitical repercussions, 174
Geothermal energy, 149
    advantages and disadvantages, 150
GHG. *see* Greenhouse gases (GHG)
Global climate change, 207–242
    current and future impacts of global warming, 219–222
    global climate policy, 228–237
    global warming causal factors and research, 211–218
    nature of problem, 207–210
    reducing greenhouse emissions, 223–227

Global climate policy, 228–237
    China, 232
Global ecosystem, 212
Global energy consumption, 130
Global footprint network, 190
Global oil production, 158, 166
Global warming
    causal factors and research, 211–218
    current and future impacts, 219–222
    policy debate, 210
    uncertainty and controversy, 208
Great climate flip-flop, 221
Greenhouse gases (GHG)
    contributions and economy, 218
    effect, 214
    emissions and ethical issues, 231
    emissions example, 46
    factors, 212
    vital statistics, 218
Gross Domestic Product (GDP)
    relationship to capita, 134
    relationship to energy consumption, 134
Growing oil consumption and lower exports by producers, 168

H

Hansen, James, 235
Hirsch Report, 169
Holian, Brian, 274
Hubbert, M. King, 164, xxvi
Hubbert Curve, 166
Hubbert Peak, 166
Hubbert's prediction, 164–167
Hydroelectric energy, 145
    advantages and disadvantages, 146
Hydrogen energy, 150, 171
    advantages and disadvantages, 152

I

Ice core evidence, 214
Implementation and operation (4.5), 59–96
    approving documents (4.5.3.2a), 72
    assessing products' energy use over time, 93
    awareness (4.5.2), 60
    awareness of energy policy (4.5.2a), 61
    benefits of improved energy performance (4.5.2c), 63
    communicating operational controls to personnel (4.5.4c), 82

Index

communications, 85
communications (4.5.5), 83–89
competence (4.5.2), 60
control (4.5.4), 78–80
controlling external documents (4.5.3.2f), 76
control of documents (4.5.3.2), 69–71
designing facilities, equipment, systems, and processes (4.5.6), 90–91
documentation requirements (4.5.3.1), 65
documenting energy policy (4.5.3.1b), 67
documenting scope and boundaries (4.5.3.1a), 66
documenting the energy policy (4.5.3.1b), 68
documents legible and identifiable (4.5.3.2e), 75
energy performance improvement, 98
energy performance improvement (4.5.2d), 64
facilities, processes, systems, and equipment (4.5.4b), 82
general (4.5.1), 59
identifying changes (4.5.3.2c), 73
maintaining documents (4.5.3.1e), 68
maintaining significant energy use (4.5.4a), 81
meaning, 59
potential disasters, 94
preventing use of obsolete documents (4.5.3.2g), 77
procurement of energy services (4.5.7.1), 92
procurement of energy supply (4.5.7.2), 95–96
relevant versions available at points of use (4.5.3.2d), 74
review and update (4.5.3.2b), 72
roles, responsibilities, and authorities (4.5.2b), 62
training (4.5.2), 60
Intergovernmental Panel on Climate Change (IPCC), 209, 210, 213, 236, 237
findings and projections, 220
Internal audit
EnMS, 100–103
meaning, 101
International stage climate change, 233

International Standards Organization 9001, 14001, and 50001 relationship between, xxi–xxiii
International Standards Organization 50001 Energy Management Systems (ISO 50001)
4.2, 5–27
4.2.1, 5–6
4.2.1a, 7
4.2.1b, 8
4.2.1c, 9
4.2.1d, 10–11, 66, 67
4.2.1e, 12
4.2.1f, 13–14
4.2.1g, 15
4.2.1h, 16
4.2.1i, 17
4.2.1j, 18
4.2.2, 19
4.2.2a, 20
4.2.2b, 21
4.2.2c, 22
4.2.2d, 23
4.2.2e, 24
4.2.2f, 25
4.2.2g, 26–28
4.3, 29–35
4.4, 37–57
4.4.1, 37–38
4.4.2, 40–41
4.4.3, 42–46
4.4.4, 48–50
4.4.5, 26, 51–52
4.4.6, 26, 53–58
4.5, 59–96
4.5.1, 59
4.5.2, 60
4.5.2a, 61
4.5.2b, 62
4.5.2c, 63
4.5.2d, 64
4.5.3.1, 65
4.5.3.1a, 66
4.5.3.1b, 67, 68
4.5.3.1e, 68
4.5.3.2, 69–71
4.5.3.2a, 72
4.5.3.2b, 72
4.5.3.2c, 73
4.5.3.2d, 74
4.5.3.2e, 75
4.5.3.2f, 76
4.5.3.2g, 77

4.5.4, 78–80
4.5.4a, 81
4.5.4b, 82
4.5.4c, 82
4.5.5, 83–89
4.5.6, 90–91
4.5.7.1, 92
4.5.7.2, 95–96
4.6, 97–116
4.6.1, 97–98
4.6.2, 99
4.6.3, 100–103
4.6.4, 104–106
4.6.4a, 107
4.6.4b, 108–110
4.6.4c, 111
4.6.4d, 112
4.6.4e, 113
4.6.5, 114
4.7, 117–125
4.7.1, 118
4.7.1a, 119
4.7.1b, 120
4.7.1c, 120
4.7.1d, 120
4.7.1e, 120
4.7.1f, 121
4.7.1g, 121
4.7.1h, 122
4.7.1i, 122
4.7.2, 123
4.7.2a, 123
4.7.2b, 124
4.7.2c, 124
4.7.2d, 124
4.7.2e, 125–126
applying, xvi
benefits of adopting, xxii
development and maintenance process, xxiv–xxv
focus, xxv
history, xix
purpose, xx
registration, 1–4
requirements, 11
system process, xxiii
IPCC. *see* Intergovernmental Panel on Climate Change (IPCC)
ISO. *see* International Standards Organization 50001 Energy Management Systems (ISO 50001)

J

Jacobson-Delucci sustainable energy plan, 200–203
Jevon's Paradox, 157

K

King Hubbert's prediction, 164–167
Kyoto Protocol, 231

L

Legible documents, 75
License renewal (LR), 274
License review process, 275
Likelihood definition matrix, 110
LR. *see* License renewal (LR)

M

Management representative
 and energy management team, 8
 responsibility, 8, 22, 24, 61
Management responsibility (4.2), 5–28
 and authority (4.2.2f), 25
 communicating importance of energy management (4.2.1e), 12
 conducting management reviews (4.2.1j), 18
 defining EnMS scope and boundaries (4.2.1d), 10–11
 determining methods for operation and control of EnMS (4.2.2g), 26–28
 energy performance indicators (4.2.1g), 15
 establish, implement, and maintain energy policy (4.2.1a), 7
 establishing energy performance objectives (4.2.1f), 13–14
 forming energy management team (4.2.2d), 23
 general requirements (4.2.1), 5–6
 improve EnMS (4.2.2a), 20
 long-term planning (4.2.1h), 16
 plan and direct energy management activities (4.2.2e), 24
 provide energy management system resources (4.2.1c), 9
 report to top management on energy performance (4.2.2c), 22

Index

report to top management on EnMS (4.2.2b), 21
representative and energy management team (4.2.1b), 8
results measured and reported (4.2.1i), 17
roles and authority (4.2.2), 19
Management review (4.7), 117–126
  actions from previous management reviews (4.7.1a), 119
  allocation of resources (4.7.2e), 125–126
  changes in energy performance of organization (4.7.2a), 123
  changes to energy policy (4.7.2b), 124
  changes to EnPIs (4.7.2c), 124
  continual improvement of EnMS (4.7.2d), 124
  corrective and preventive actions (4.7.1g), 121
  energy management system audit results (4.7.1f), 121
  energy objectives and targets (4.7.1e), 120
  energy performance and indicators (4.7.1c), 120
  energy policy (4.7.1b), 120
  input (4.7.1), 118
  legal compliance (4.7.1d), 120
  output (4.7.2), 123
  projected energy performance (4.7.1h), 122
  recommendations for improvement (4.7.1i), 122
  summary of requirements, 117
Mann, Michael, 236
Methane, 219
Methods, tools and techniques, 265–272
  advantages and disadvantages, 271
  applying NGT to maximize return on investment, 269
  typical NGT procedure, 270
Microfinance, 196
Mitigation, 232
Municipal solid waste, 149

N

National Environmental Policy Act-U.S. (NEPA), 41
Nation's population base *vs.* energy consumption, 135
Natural gas combined-cycle generation, 141

NEPA. *see* National Environmental Policy Act-U.S. (NEPA)
Nominal Group Technique (NGT) process, 269
  methods, tools and techniques, 270
  stages, 270
Nonconformity risk decision guide, 111
Nonconventional petroleum, 171
NRC. *see* Nuclear Regulatory Commission (NRC)
Nuclear, 141, 172
Nuclear fission, 141
Nuclear Regulatory Commission (NRC) license review process, 274, 275

O

Oceanic Conveyor Belt, 222
One-child carbon offset credit policy of China, 232
OPEC. *see* Organization of the Petroleum Exporting Countries (OPEC)
Operational controls, 78–80
  communicating to personnel, 82
Operations
  and activities trends towards energy significance, 79
  significant energy use, 79
Organization, 243–244
  accomplish all energy activities, 23
  operations and activities trends towards energy significance, 79
  qualifications, 2
Organization of the Petroleum Exporting Countries (OPEC), 165–176
Outputs of energy review, 97

P

Paleoclimatology, 214
Peaking of World Oil Production, 169
Peak Oil, xxvi, xxvii
  assessment, xxvii
  growing oil consumption and lower exports by producers, 168
  implications, 176
  King Hubbert's prediction, 164–167
  looming crisis, 163–177
  policy implications, 169–175
  preparation, 174
  solution, 168

theory, 164
US department of energy's assessment, 168
Performance. *see* Checking performance
Periodic review and update, 72
Perspectives on energy efficiency and conservation, 155–162
  issues, 160
  worldwide, 158–159
Pickens sustainable energy plan, 198–200
Piltdown Man, 236
Pimentel, David, 176
Planning
  conceptual overview, 39
  energy significant operations eight steps, 80
  sustainable development, 181
Policy integration, 181
Precautionary principle, 181, 235
Preengineering decisions for baseline, 47
Preventive actions defined, 105
Private equity and venture capital, 195
Proactive personnel action communications, 87
Procedures awareness, 61
Procurement
  defined, 95
  energy services, products, and equipment, 92
  energy supply, 95–96
Products' energy use over time assessment, 93
Professional organization standards, 41
Project finance, 195

R

Records
  control, 114
  list, 115
  maintaining energy review, 42
Reducing greenhouse emissions, 223–227
Relevant versions, 78
Renewable energy, 136
  sources, 131, 144
  technology potential, 223
Re-review of license renewal, 278
Reserve estimates, 166
Resource development applications, 193–195
Review
  consider, 91

  and determining causes of nonconformities, 107
  effectiveness of action taken, 113
Risk-consequences scoring term, 110

S

SG. *see* Smart grid (SG)
Significant energy use, 244
  apply, 97
  operations, 79
Smart grid (SG), 136
Socially responsible investments, 195
Solar energy, 146
  advantages and disadvantages, 147
  variation, 217
Steinhart, John, 176
Stommel, Henry, 222
Survey of energy sources, 139–152
Suspended fire hose method, 226
Sustainability and energy policy, 179–205
  achieving, 181
  Agenda 21, 184
  applications of sustainable resource development, 193–195
  case studies examples, 192
  common principles of sustainability, 184–188
  definitions of sustainability, 179–181
  dilemmas, 189
  Dow Jones measures, 194
  economic costs, 188
  economy and Worldwatch innovations, 195
  energy consumption, generation, sustainability, and energy systems, 138
  examples of sustainable energy policy, 196–202
  hierarchy, 188
  impact, 182–183
  measuring sustainable development, 189–192
  principles, 184–188
  sustainability impact, 182–183
  US executive order, 186

T

Techniques. *see* Methods, tools and techniques
Technology transfer, 232

Index

Temperature and carbon dioxide correlation, 216, 217
Three Mile Island, 273
Three-step energy review process, 45
Tidal energy, 151
Tools. *see* Methods, tools and techniques
Top management
    appoint energy management representative, 19
    commitment and energy management, 12
    express organization's overall intentions, 7
    responsible for organization's EnMS, 20
Trade wars, 234
Training, 60

U

United States
    department of energy's assessment, 168
    global change research program, 221
    National Environmental Policy Act (NEPA), 277
    NRC's Division of License Renewal (DLR), 274
    oil production, 165

Wadman-Markey cap-and-trade bill, 229
Uranium-235 atoms, 141

V

Venture capital, 195

W

Wadman-Markey cap-and-trade bill, 229. *see also* Cap-and-trade policy
Wave and ocean energy, 149
Whitening clouds, 226
Wind energy, 145
    advantages and disadvantages, 145
    tidal, solar, and geothermal, 172
Wood waste, 148
World energy consumption
    energy use and efficiency, 159
    industrial sector, 133
    versus nation's population base, 135
    projected increase, 132
World marketed energy, 133
World oil production developments, 165
World peak, xxvi–xxvii
World's first energy crisis, xxxi
Worldwatch innovations, 195
Written training, 64

	DATE DUE		